Preservation of Arsenic Species

Awwa
Research
Foundation

About the Awwa Research Foundation

The Awwa Research Foundation (AwwaRF) is a member-supported, international, nonprofit organization that sponsors research to enable water utilities, public health agencies, and other professionals to provide safe and affordable drinking water to consumers.

The Foundation's mission is to advance the science of water to improve the quality of life. To achieve this mission, the Foundation sponsors studies on all aspects of drinking water, including supply and resources, treatment, monitoring and analysis, distribution, management, and health effects. Funding for research is provided primarily by subscription payments from approximately 1,000 utilities, consulting firms, and manufacturers in North America and abroad. Additional funding comes from collaborative partnerships with other national and international organizations, allowing for resources to be leveraged, expertise to be shared, and broad-based knowledge to be developed and disseminated. Government funding serves as a third source of research dollars.

From its headquarters in Denver, Colorado, the Foundation's staff directs and supports the efforts of more than 800 volunteers who serve on the board of trustees and various committees. These volunteers represent many facets of the water industry, and contribute their expertise to select and monitor research studies that benefit the entire drinking water community.

The results of research are disseminated through a number of channels, including reports, the Web site, conferences, and periodicals.

For subscribers, the Foundation serves as a cooperative program in which water suppliers unite to pool their resources. By applying Foundation research findings, these water suppliers can save substantial costs and stay on the leading edge of drinking water science and technology. Since its inception, AwwaRF has supplied the water community with more than $300 million in applied research.

More information about the Foundation and how to become a subscriber is available on the Web at **www.awwarf.org.**

Preservation of Arsenic Species

Prepared by:
Gautam Samanta and **Dennis A. Clifford**
The University of Houston
Department of Civil and Environmental Engineering
Houston, TX 77204-4791

Sponsored by:
Awwa Research Foundation
6666 West Quincy Avenue, Denver, CO 80235-3098

Published by:

Awwa Research Foundation

American Water Works Association

IWA Publishing

DISCLAIMER

This study was funded by the Awwa Research Foundation (AwwaRF). AwwaRF assumes no responsibility for the content of the research study reported in this publication or for the opinions or statements of fact expressed in the report. The mention of trade names for commercial products does not represent or imply the approval or endorsement of AwwaRF. This report is presented solely for informational purposes.

CONTENTS

* Note: Chapter 5 is a stand-alone report on the effect of sulfide on arsenic preservation and speciation. It describes the Project Continuation Request (PCR) research completed following the submission of the draft report on this project, i.e., project 2815 — Preservation of Arsenic Species.

TABLES

FIGURES

FOREWORD

The Awwa Research Foundation is a nonprofit corporation that is dedicated to the implementation of a research effort to help utilities respond to regulatory requirements and traditional high-priority concerns of the industry. The research agenda is developed through a process of consultation with subscribers and drinking water professionals. Under the umbrella of a strategic Research Plan, the Research Advisory Council prioritizes the suggested projects based upon current and future needs, applicability, and past work; the recommendations are forwarded to the Board of Trustees for final selection. The foundation also sponsors research projects through the unsolicited proposal process; the Collaborative Research, Research Applications, and Tailored Collaboration programs; and various joint research effects with organizations such as the U.S. Environmental Protection Agency, the U.S. Bureau of Reclamation, and the Association of California Water Agencies.

This publication is a result of one of these sponsored studies, and it is hoped that its findings will be applied in communities throughout the world. The following report serves not only as a means of communicating the results of the water industry's centralized research program but also as a tool to enlist the further support of the nonmember utilities and individuals.

Projects are managed closely from their inception to the final report by the foundation's staff and large cadre of volunteers who willingly contribute their time and expertise. The foundation serves a planning and management function and awards contracts to other institutions such as water utilities, universities, and engineering firms. The funding for this research effort comes primarily from the Subscription Program, through which water utilities subscribe to the research program and make an annual payment proportionate to the volume of water they deliver and consultants and manufacturers subscribe based on their annual billings. The program offers a cost-effective and fair method for funding research in the public interest.

A broad spectrum of water supply issues is addressed by the foundation's research agenda: resources, treatment and operations, distribution and storage, water quality and analysis, toxicology, economics, and management. The ultimate purpose of the coordinated effort is to assist water suppliers to provide the highest possible quality of water economically and reliably. The true benefit are realized when the results are implemented at the utility level. The foundation's trustees are pleased to offer this publication as contribution to wards that end.

Walter J. Bishop
Chair, Board of Trustees
Awwa Research Foundation

Robert C. Renner, P.E.
Executive Director
Awwa Research Foundation

ACKNOWLEDGMENTS

The authors are indebted to the following persons for their cooperation and participation in this project:

1. Melinda (Mindy) Erickson, a doctoral student of Prof. Randal Barnes in the Department of Civil Engineering at the University of Minnesota for collecting, preserving, and field speciating the samples from 48 different Minnesota drinking water wells
2. Dr. David Chang, Water Quality Manager, Region II, the Southern California Water Company, California for arranging and supervising the collection, preservation, and field speciation of the California drinking water wells.
3. Mr. Craig Lacher, Lab Supervisor of the City of Grand Forks, North Dakota Drinking Water Laboratory for collecting, preserving, and field speciating the North Dakota drinking water well samples.
4. Dr. Rick A. Knurr, Senior Scientist, Department of Geology and Geophysics, University of Minnesota for analyzing the Minnesota well waters for total arsenic using ICP-MS and for participating in a UH/UMinn interlaboratory comparison of total arsenic results by ICP-MS and hydride generation AA.
5. Project Advisory Committee members Dr. Andy Eaton, Prof. Janet Hering, and David Kimbrough, for their many useful suggestions and observations throughout the project.
6. AwwaRF Project Officer, Dr. Djanette Khiari for her valuable administrative support and the many helpful technical suggestions she provided.

The authors are also grateful for the funding provided by Awwa Research Foundation and for the in-kind matching funding provided by the University of Houston, the University of Minnesota, the Southern California Water Company, and the City of Grand Forks North Dakota.

EXECUTIVE SUMMARY

In arsenic-contaminated ground waters, only inorganic arsenic in the form of arsenite [As(III)] and arsenate [As(V)] are significant. Concentrations of organic forms are generally low or negligible in groundwater. However, the distribution of As(III) and As(V) species depends greatly on the abundance of redox-active solids, specially organic carbon, the activity of microorganisms, and the extent of diffusion of O_2 from the atmosphere. In strongly reducing aquifers, As(III) is the dominant species based on the thermodynamic considerations, whereas As(V) is the more stable oxidation state under oxic conditions or in oxygenated waters. Although the availability of As(V) is limited by adsorption onto hydrated ferric oxides (HFOs), hydrous aluminum and manganese oxides and clays, As(III) may also adsorb onto these solids.

It is important to know the actual concentrations of As(III) and As(V) in drinking waters to establish the treatment strategy to remove arsenic. Due to the anionic character of As(V), it can be removed more easily than As(III). Again, due to wide variation of toxicity of As(III) and As(V), epidemiologists should know the actual concentration of each arsenic species in the drinking water.

Unfortunately, arsenic species are not stable in water samples, and the distribution of arsenic species can change during transportation and storage. Normally, the time between on-site sampling and arsenic analysis in the laboratory is at least 1 day, and, within this period, significant changes water chemistry and arsenic speciation can take place. So, it is important to stabilize the distribution of arsenic species at the moment of sampling, and to preserve the original speciation until the analysis is done, especially when iron is present in the sample.

OBJECTIVES

The main objective of this study was to develop a method to preserve inorganic arsenic species As(III)/As(V) in water samples for at least one week and possibly 28 days in order to allow speciation and analysis at a remote laboratory. To attain the main objective, the following specific tasks were carried out:

1. Identify existing preservation methods that show potential for adaptation to preserve inorganic arsenic species in water samples.
2. Compare the relative advantages and disadvantages of the identified preservation methods
3. Evaluate the effect of ferrous iron on the identified speciation methods.
4. Study the effect of UV light on the preserved samples.
5. Study the effect of initial redox potential (Eh) and pH on the stability of arsenic species in the preserved samples.
6. Develop the best preservation method and evaluate it for its preservation efficiency.
7. Develop a laboratory speciation method for inorganic species.
8. Modify the field speciation kit to cut costs and reduce sample size.
9. Verify the performance of the best preservation method for field samples by comparing field speciation to lab speciation of preserved samples over a period of 30-60 days.
10. Establish the effects of sulfide on the EDTA-HAc preservation-speciation process.

APPROACH

The project included an information-gathering phase in addition to field and laboratory testing. The literature review provided the necessary information and an up-to-date research summary on the preservation of inorganic arsenic species. Based on the review, it was obvious that the preservation of arsenic species in water samples is controversial. Furthermore, no study had used a comprehensive approach to possible interferences. A few studies were conducted in the laboratory by using insoluble Fe(III) to observe the effect of iron on arsenic species distribution. However, iron is mostly present in the soluble ferrous [Fe(II)] state in ground water. Therefore, it was more reasonable to study the effect of ferrous iron on the stability of arsenic species. So, a comprehensive, systematic study was conducted in our laboratory. Among the promising preservatives described in the literature, H_2SO_4, H_3PO_4, EDTA, and EDTA-HAc were chosen for the systematic study, which is outlined below:

1. Prepare a synthetic groundwater (NSF-2 challenge water) whose composition is similar in anionic composition to arsenic-contaminated USA ground waters used for drinking water supply.
2. Prepare anoxic, low-DO synthetic groundwater and spike it with ferrous iron in order to study the influence of Fe(II) on the arsenic species preservation process.
3. Study in effectiveness of H_2SO_4, H_3PO_4, EDTA, and EDTA-HAc in the presence and absence of ferrous iron.
4. Study the effect of redox potential (Eh), initial pH and UV light on the stability of preserved and unpreserved arsenic species.
5. Study the influence of the chosen preservation method on the stability of arsenic species in field samples from cooperating utilities.
6. Study the influence of sulfide on the preservation and speciation of arsenic(III)/(V).

CONCLUSIONS

This study of the preservation of inorganic arsenic species-10μg/L As(III) and 10 μg/L As(V) in a representative challenge water-resulted in the following conclusions: (Unless otherwise stated, all experiments were carried out at room temperature under laboratory light and successful preservation meant less than 10% change in As(III) concentration.)

1. In challenge water without iron, all three preservatives-H_2SO_4, H_3PO_4, and EDTA-HAc-could could preserve the arsenic species for at least 7 days, the maximum duration of the test. Without preservatives, the arsenic species were stable for only one day.
2. Contrary to reports in the literature, EDTA alone could not preserve the arsenic species for more than twenty-four hours even in the absence of Fe(II).
3. In challenge water containing 3 mg/L Fe(II), all three preservatives-H_2SO_4, H_3PO_4, and EDTA-HAc-could preserve the arsenic species for a week or more in opaque brown polypropylene bottles under laboratory light. H_3PO_4 was an effective preservative for seven days, while H_2SO_4 and EDTA-HAc could preserve the arsenic species for at least 28 days without significant changes in their concentrations.
4. When the same experiments with challenge water containing iron were carried out in crystal clear PET bottles under laboratory light, H_3PO_4 and H_2SO_4 could preserve the

speciation for only 7 days while EDTA-HAc preservation was successful for 21 days. Thus, normal laboratory lighting was detrimental to preserving speciation in crystal clear plastic bottles with all three preservatives.

5. Under strong UV-A illumination (simulated direct sunlight) in the presence of Fe(II), none of the preservatives-EDTA-HAc, H_2SO_4, or H_3PO_4-could stabilize the arsenic species in clear, transparent plastic bottles for more than one hour. Under these extreme conditions, H_3PO_4 was a much better preservative than EDTA-HAc or H_2SO_4. Based on the results of the experiments with normal laboratory light and simulated direct sunlight, it is strongly recommended that samples be preserved in opaque bottles during field sample collection, transportation, and storage.
6. Under reducing conditions (Eh = -100 ± 10 mV) in the pH range of 6.5 to 8.4, and in the presence of Fe(II), EDTA-HAc could preserve the As species up to 28 days, whereas H_2SO_4 was effective for only 21 days and H_3PO_4 was effective for only 3 days at pH 6.5. Without preservatives, the speciation was immediately changed under reducing conditions.
7. Under oxidizing conditions (Eh = +200 ± 10 mV) and in the presence of >0.2 ppm DO, H_2SO_4 and H_3PO_4 could not preserve the arsenic species for more than a few minutes in opaque brown bottles. However, EDTA-HAc could preserve the arsenic species under these oxidizing conditions for at least 28 days. Some changes in speciation did occur with EDTA-HAc, but they were within the allowable 10% deviation.
8. Based on extensive experimental results in Fe(II)-contaminated challenge water, it was concluded that EDTA-HAc could be used to preserve the arsenic species for at least 28 days in opaque plastic bottles. Thus, EDTA-HAc was chosen as the preservation method for further testing. Although the alternative preservatives, H_2SO_4 and H_3PO_4, successfully preserved the original As(III)/(V) speciation under some conditions, these preservatives were generally unsuccessful for the desired 28-day period under reducing and oxidizing conditions in the sample pH range of 6.5-8.4 and in the presence of 3 mg/L Fe(II).
9. Verification tests of the EDTA-HAc preservation method were conducted at field locations in Minnesota, California, and North Dakota. Field samples were collected in bottles containing EDTA-HAc preservative and immediately speciated using the EDTA-HAc field-speciation method developed at UH under AwwaRF Project 2663. Preserved and speciated samples were shipped without refrigeration to the University of Houston by overnight mail and received within 54 hours of sampling. The field-speciated samples were analyzed for As(III) by FI-HG-AAS and the results were compared with the lab-speciated As(III) results using differential pH FI-HG-AAS on the EDTA-HAc preserved samples. These two data sets were highly correlated with a slope near 1.0 (slope 0.977; R^2 = 0.9986), which indicates that during transportation (27–55 h) the As(III)/(V) distribution in the EDTA-HAc preserved samples did not change. Further statistical analysis using the paired t-test showed no significant difference between the methods at the 95% confidence level. In the laboratory all the samples were preserved for at least 30 days and the results showed that except for one sample, all the preserved samples maintained their species distribution for at least 30 days. The triplicate samples were monitored for a total of 85 days, and no statistically significant changes were observed.

10. Because ICP-MS is so widely used for arsenic analysis-in spite of its well known chloride interference, the ICP-MS method was compared with FI-HG-AAS in this study. The ICP-MS and FI-HG-AAS data were highly correlated (R^2 = 0.9932), however, slope of the plot was significantly different than 1.0. (Slope = 1.08x), i.e., the ICP-MS data were consistently about 8% lower that FI-HG-AAS data. So, for routine analysis FI-HG-AAS is the more useful because it is requires less capital investment and is subject to less serious interference.
11. The main conclusion from the study is that the EDTA-HAc preservation method is capable of preserving inorganic As(III)/(V) speciation for more than 30 days in well waters under reducing or oxidizing conditions with or without ferrous iron present. Additionally, samples preserved with EDTA-HAc can be speciated immediately in the field using ion-exchange minicolumns containing strong-base anion resin in the chloride form. Results from immediate As(III)/(V) field speciation compared with lab speciation up to 30 days later are not statistically different.
12. Based on the literature and as confirmed in our experiments, the presence of sulfide and arsenic together leads to very complicated thioarsenite chemistry. EDTA-HAc could not effectively preserve the arsenic species when sulfide was present. (However, when sulfide is present in an arsenic-contaminated water, As(III)/(V) speciation is unnecessary because As(V) cannot co-exist with sulfide.) When sulfide concentration was high (>1.0 mg/L) precipitation of As_2S_3(am) occurred, and only As(Tot) could be measured in preserved samples after the treatment with NaOCl. As(III) and thioarsenite species could be speciated by anion-exchange resin. The best way to measure As(III) and thioarsenite species was to speciate them immediately after collection of sample without any preservative in the field by anion-exchange resin. The difference between As(Tot) and As(III) in the effluent gave the value of thioarsenite species.

RECOMMENDATIONS

Based on the theoretical and experimental results from this project, the following recommendations can be made relative to the analysis and speciation of inorganic arsenic (III)/(V) in ground waters with and without sulfide present:

1. Arsenic-containing ground waters that do not contain detectable levels of sulfide may be preserved with EDTA-HAc and immediately speciated in the field with chloride-form resin minicolumns. Samples preserved in this way may be stored at room temperature and analyzed up to 30 days later for As(Tot) and/or speciated again for As(III)/(V) using ion exchange or differential pH hydride generation in the laboratory. For these samples As(V) is retained on the columns and As(III) passes through the column as H_3AsO_3. The relevant equations for non-sulfidic waters are as follows:

$$As(Tot) = As(III) + As(V)$$
$$As(V) = As(Tot) - As(III)$$

2. If an arsenic-contaminated ground water to be speciated for As(III)/(V) contains detectable sulfide, it should not be preserved with EDTA-HAc or any mineral acid.

Acidifying such a sample will likely precipitate As_2S_3(am) and cause falsely low As(III) and As(Tot) analyses. The unpreserved sample may be analyzed later for As(Tot) by adding NaOCl (0.02-0.1 M) to oxidize and redissolve arsenic sulfides prior to As(Tot) analysis by hydride generation AAS or alternative method.

3. Sulfide-containing ground waters will contain only As(III) in the form of thioarsenite species and H_3AsO_3, and may be immediately speciated in the field (without preservation) using chloride-form anion resin minicolumns to separate thioarsenite species from H_3AsO_3. The quantification of thioarsenite species may be important because of their lower bioavailability and reduced toxicity compared with arsenite (H_3AsO_3) species.
4. Thioarsenite species concentration may be determined as the difference between As(Tot) and As(III) as measured in the IX eluent of a chloride-form anion column used to speciate an unpreserved sulfidic ground water. The relevant equation are as follows:

$$\mathrm{As(Tot)} = \mathrm{As(Thioarsenite)} + \mathrm{As}(H_3AsO_3)$$
$$\mathrm{As(Thioarsenite)} = \mathrm{As(Tot)} - \mathrm{As}(H_3AsO_3)$$

The EDTA-HAc methods should be further verified. The combined results of two AwwaRF arsenic analysis projects — No. 2663: Field Speciation Method for Arsenic Inorganic Species and this project: No. 2815 Preservation of Arsenic Inorganic Species — have yielded a method for immediate field-speciation and a preservation technique for speciation within 30 days of sampling. Limited field verifications of these methods have been completed, which demonstrate that they give statistically indistinguishable results. What is needed now is an inter-laboratory study of the field-speciation and preservation methods to further validate their accuracy and precision. An inter-laboratory study is recommended and it should include the following tasks: (1) solicitation of 3-5 analytical labs and cooperating utilities, (2) selection of a central lab to prepare EDTA-HAc preserved As(III)/(V) standards and send them to the cooperating labs using different methods of arsenic analysis, (3) collection and analysis of accuracy and precision data on the standards, (4) collection, field speciation, and preservation of representative field samples, (5) analysis of speciated and preserved samples by the cooperating labs, (6) collection and analysis of data on the field-speciated and preserved samples, and (7) preparation of a final report.

CHAPTER 1
INTRODUCTION

OVERVIEW

Inorganic arsenic is considered as a human carcinogen with multiple sites of attack. Epidemiological studies have demonstrated the higher risks of skin, bladder, lung, liver and kidney cancer along with other non-cancerous health effects that result from continued consumption of elevated levels of As in drinking water (Guha Mazumder at al. 1998, Chen et al. 1988, Ferreccio et al. 2000). Due to the elevated health risk, based on the analysis of the United States Environmental Protection Agency (USEPA) and two independent reports by the National Research Council (NRC), USEPA has reduced the maximum contamination level (MCL) of arsenic in drinking water from 50 μg/L to 10 μg/L (US EPA 2001).

Arsenic is widely distributed element to which humans are exposed through ingestion of water and food, and by inhalation. Many countries around the world are exposed to elevated levels of arsenic in their drinking water. But, the most devastating incidence of arsenic poisoning has been reported in Bangladesh, and West Bengal-India (Chowdhury et al. 2000, Chowdhury et al. 2000a, Das et al. 1995). Reportedly, more than 6 million people in West Bengal, India, and more than 70 million people in Bangladesh are drinking ground water containing elevated levels (50 μg/L and above) of arsenic (Chakraborti et al. 2002). In Bangladesh out of an estimated 6-11 million shallow tube wells, ~ 27% are contaminated with arsenic above 50 μg/L (Kinniburgh and Kosmus 2002). Here in the USA, according to the US Geological Survey (USGS) and the USEPA, high concentrations of arsenic are widespread in Western, Midwestern and Northwestern United States (http://co.water.usgs.gov/trace/arsenic). The EPA has reported that more than 5.5% of the total water supply systems in USA contain arsenic at a level greater than the present limit (10 μg/L). Higher levels of arsenic tend to be found more often in groundwater than in surface water sources. About 3,000 of the nation's 54,000 Community Water Supplies and 1,100 of the 20,000 Non-Transient Non-Community water Supplies exceed the current 10 μg/L limit. According to the Natural Resources Defense Council (NRDC, 2000), over 34 million Americans drink water that increases their arsenic-related cancer.

The occurrence of elevated levels of arsenic in groundwater is due to the various geochemical reactions including oxidation of arsenic-rich pyrite (Roy Chowdhury et al. 1999), reductive dissolution of arsenic-rich iron hydroxides (Nickson et al. 1998), and bicarbonate (HCO_3^-) leaching of arsenic from sulfide minerals under anaerobic conditions (Kim and Nriagu 2000). Aquatic arsenic chemistry is very complex due to the numerous reaction possibilities and the multiple oxidation states of arsenic. In ground waters, concentrations of organic forms are generally low or negligible, and only inorganic arsenic in the form of arsenite [As(III)] and arsenate [As(V)] are significant (Watt and Le 2003, Irgolic 1994). However, the concentration of As(III) and As(V) depends greatly on the abundance of redox-active solids (especially organic carbon), the activity of microorganisms, and the extent of diffusion of O_2 from the atmosphere (Smedley and Kinniburgh 2002). In strongly reducing aquifers, As(III) is the dominant species as expected from the thermodynamic considerations whereas As(V) is the more stable oxidation state under oxic condition. Although the presence of soluble As (V) is limited by adsorption onto hydrated ferric oxides (HFOs), aluminum hydroxide, manganese oxides, and clays (Belzile and Tessier 1990, Farmer and Lovell 1986, Takamatsu et al. 1985), As(III) may also adsorb onto these solids. In sulfide-rich waters, As(III) solubility is limited by the formation of arsenic sulfide (Webster 1990) or arsenic bearing iron sulfides (Moore et al. 1988).

It is important to know the actual concentrations of As(III) and As(V) in drinking waters to establish the treatment strategy to remove arsenic. Due to the anionic character of As(V), it can be removed more easily than As(III). Again, due to wide variation of toxicity of As(III) and As(V), epidemiologists should know the actual concentration of each arsenic species in the drinking water.

Unfortunately, arsenic species are not stable in water samples, and the distribution of arsenic species changes during transportation and storage. The major causes of arsenic species changes are (1) precipitation of hydrated ferric oxides (HFOs) accompanied by differential adsorption of arsenic species (Pierce and Moore 1982), (2) microbial activity (Meng et al. 2003), (3) abiotic oxidation of As(III) specifically, in the presence of acids (Hall et al. 1999, Bednar et al. 2002), and (4) photo-oxidation in the presence of ultraviolet radiation (Hug et al 2001, Lee and Choi 2002, Emett and Khoe 2001). The commonly recommended strategies to preserve the arsenic species include filtration, acidification, cooling, storage in the dark, addition of complexing agents such as EDTA, and minimization of storage time.

Filtration is standard practice in ground water sampling for arsenic. Filtration can suppress the conversion of arsenic species partly by removing HFOs and partly through the removal of some bacteria. Wilkie and Hering (1998) reported rapid changes in arsenic distribution in unfiltered samples.

Acidification is a widely used protocol to preserve the arsenic species in water samples. Acidification reduces the precipitation of HFOs by increasing their solubility. HCl, HNO_3, H_2SO_4, H_3PO_4, and ascorbic acid were used in the arsenic preservation studies of (Hall et al. (1999), Bednar et al. (2002), Edwards et al. (1998), and Daus et al. (2002).

Refrigeration could preserve the arsenic species (Hall et al. 1999, Palacios et al. 1997) presumably by slowing down the microbial activities and keeping the sample away from direct sunlight. But, the analysis of well water in Genesee County, Michigan, showed the oxidation of As(III) to As(V) within three days of storage in a refrigerator (Kim 2001). Aggett and Kriegman (1987) and Daus et al (2002) also reported the loss of arsenic in freezing samples.

Many researchers reported the preservation problem of arsenic species in iron-rich waters. Gallagher et al (2001) and Bednar et al (2002) demonstrated that EDTA was an effective preservative for arsenic species in Fe(III)-rich water samples. According to Bednar et al, EDTA alone could preserve the arsenic species without lowering the pH of the sample in opaque bottles. On the other hand, Gallagher et al. (2001) reported the need to lower pH to about 3.2 with acetic acid to preserve the arsenic species using EDTA.

OBJECTIVES

The main objective of this study was to develop a method to preserve inorganic arsenic species As(III)/As(V) in water samples for at least one week and preferably 28 days in order to allow speciation and analysis at a remote laboratory. To meet the main objective, the following specific tasks were carried out:

Identify existing preservation methods that show potential for adaptation to preserve inorganic arsenic species in water samples.

1. Compare the relative advantages and disadvantages of the identified preservation methods
2. Evaluate the effect of ferrous iron on the identified speciation methods.
3. Study the effect of UV light on the preserved samples.
4. Study the effect of initial redox potential (Eh) and pH on the stability of arsenic species in the preserved samples.
5. Develop the best preservation method and evaluate it for its preservation efficiency.
6. Develop a laboratory speciation method for inorganic species.
7. Modify the field speciation kit to cut costs and reduce sample size.
8. Verify the performance of the best preservation method for field samples by comparing field speciation to lab speciation of preserved samples over a period of 30-60 days.
9. Establish the influence of sulfide on the EDTA-HAc preservation and speciation methods.

SIGNIFICANCE OF THE PROJECT

Redox potential, pH, matrix composition, radiation and flux of incident light etc. often influence the stability of arsenic species in environmental samples. Because toxicity and removal methods depend on As(III)/As(V), it is important to know the distribution of arsenic species at the point-of-sources. Normally, the time between on-site sampling and arsenic analysis in the laboratory is 1 day or longer, and during this period significant changes in the chemical composition may take place. Consequently, it is important to preserve the distribution of arsenic species at the moment of sampling until the arsenic speciation analysis is complete. Preservation is especially important when iron is present in the sample and the presence of sulfide is also expected to influence the preservation and speciation methods.

REVIEW OF PREVIOUS STUDIES

Arsenic species tend to be unstable in natural water depending on the background ions, redox condition, the presence of precipitating metals such as Fe(II), and microbial activity. Several laboratory-based analytical methods are available to accurately determine the distribution of As(III) and As(V) in water samples. But, the ability to preserve the native (point-of-sampling) distribution until the sample can be analyzed in the laboratory is, to date, an unsolved problem. The literature on the stability of the inorganic arsenic species, As(III) and As(V), in the aquatic environment is controversial. Hall et al. (1999) and Van Elteren et al. (1991) reported that As(V) slowly reduced to As(III) in de-ionized water unless water was stored at lower temperature. The reduction was faster at the lower concentration of As(V). As(V) could also be reduced in some natural waters even after a sample was filtered to remove living organisms (Hall et al 1999, Val Elteren et al. 1991).

Using a ^{73}As-radiotracer technique, Van Elteren et al. (1991) reported that the gradual oxidation of As(III) in natural and fresh water samples could be eliminated by lowering the pH to about 2. Cherry et al. (1979) reported that, in the presence of the common redox reagents in natural waters, such as O_2, H_2O, and Fe, the rates of redox reactions of inorganic arsenic were sufficiently slow to enable the water samples to be analyzed in the laboratory after transportation. Tallman and Shaikh (1980) supported the observations of Cherry et al. Other researchers made attempts to stabilize the inorganic arsenic species by lowering pH and temperature. Feldman (1979) demonstrated that ascorbic acid could be able to preserve the arsenic species over an 18 d period. However, in a study of the oxidation state of arsenic in coal ash leachates, Turner (1981) reported that 0.1% ascorbic acid was unsatisfactory to prevent the oxidation of As(III). About 75% of As(III) originally present was oxidized after storage for 33 d. Edwards et al. (1998) also reported that addition of 0.1% ascorbic acid or 0.05 M hydrochloric acid was not effective to preserve the arsenic species in natural waters.

Cheam and Agemian (1980) studied the preservation of inorganic arsenic in de-ionized and natural waters at three pH levels (1.5, 5.4, and 7.2) in different types of containers and concluded that arsenic species could be preserved in both de-ionized and natural water at room temperature in polyethylene or pyrex bottles with 0.2% (v/v) sulphuric acid. It should be mentioned here that all the experiments were carried out spiking As(III) and As(V) separately and arsenic was measured as total arsenic. So, the conversion from one state to other was not measured.

Edwards et al. (1998) used 0.05 % H_2SO_4 as an effective preservative for the inorganic arsenic species. Van Elteren et al. (1991) and Al-Sibaai and Fogg (1973) observed the oxidation of As(III) in a glass container. Again, Hall et al. (1999) reported that oxidation of As(III) to As(V) could not be retarded in the presence of 0.1% nitric or hydrochloric acid. Many researchers reported the preservation problem of arsenic species in iron-rich waters. According to Aggett and Kriegman (1987), samples should be preserved at pH 2.0 to hold iron in solution, otherwise iron could be precipitated and arsenic would be co-precipitated with iron. They also recommended preserving samples at lower temperature (0 °C) without de-oxygenation. Borho and Wilderer (1997) concluded that through the addition of HCl ($pH < 2$) and a great surplus of Fe(II), the As(III) concentration could be kept constant even in oxygen-saturated Fe(III)-containing samples for at least one week. Daus et al (2001) found the most satisfactory result for the preservation of arsenic species in iron rich water samples with 0.01M H_3PO_4 at 6 °C. Bednar et al. (2002) evaluated the efficiencies of H_2SO_4, HNO_3, HCl, and EDTA as preservatives for arsenic-containing reagent waters, with and without Fe(III) (1 mg/L) in the presence or absence of ambient light at room temperature. They reported that in the presence of Fe(III), H_2SO_4, HNO_3, and HCl could not maintain the distribution of arsenic species in ambient light or in the dark. EDTA alone could preserve the arsenic species without lowering the pH of the sample in opaque bottles. In contrast, Gallagher et al. (2001) reported that pH adjustment to 3.2 using acetic acid was necessary to preserve the arsenic species using EDTA.

In most of the laboratory tests described above, reagent grade water without typical background contaminants was used. Furthermore, only Fe(III) was used to establish the influence of iron on arsenic species preservation. It is more reasonable to use Fe(II) because preservation should prevent its oxidation to Fe(III), which is not normally present in ground water because of its insolubility.

There are also conflicting reports about the stability of arsenic species in the presence of Fe(III)(hydr)oxides. De Vitre et al (1991) reported oxidation of As(III) within 2 d irrespective of pH, while Oscarson et al. (1981) and Hug et al. (2001) reported no oxidation of As(III) in the presence of Fe(III)(hydr)oxides over a long period of time in the dark. But the oxidation of As(III) in the dark in the presence of Fe(II) was very fast at neutral pH. After aeration of anoxic groundwater containing Fe(II) the composition of the water changes rapidly. According to Stumm and Morgan (1996), at ground water pH values in the range of 7.2–8.0, Fe(II) was >90% oxidized and precipitated in the form of Fe(III)(hydr)oxides within seconds to ten minutes upon exposure to air. The presence of a redox couple, e.g. Fe(II)/Fe(III), and the pH are very important parameters to control the arsenic composition in water samples. Again, the photochemical degradation of organic compounds such as citrate (Hug et al. 2001), oxalate (Balmer and Sulzberger 1999, Sulzberger 1992), EDTA (Karametaxas et al. 1995, Litter et al. 1991) with initially Fe(III) or in suspension of hematite or lepidocrocite provides the most efficient pathway for the oxidation of As(III) and other elements at pH values between 5 and 8.

CHAPTER 2
MATERIALS AND METHODS

REAGENTS

All reagents used were of analytical reagent grade. Milli Q water (18 Ω) was used to prepare all solutions. Primary standards of 100 mg As/L of each species were prepared from sodium m-arsenite ($NaAsO_2$,) for As(III) and Sodium arsenate for As(V) (Na_2HAsO_4) both from Sigma Chemical Co, MO. Working standard solutions were prepared daily with proper dilution. Citric/citrate buffer solution was prepared using 2 M citric acid and pH was adjusted to 5.0 using NaOH. A 4-mg/mL solution of L-cysteine (Sigma Chemical Co, MO) in HCl solution was used to reduce As(V) to As(III). Sodium tetrahydroborate (EM Science, Germany) solutions were prepared fresh daily, and were supplemented with sodium hydroxide.

To prepare challenge water, the following salts were used $NaNO_3$, $NaHCO_3$, $Na_2HPO_4 \cdot H_2O$, NaF, $Na_2SiO_3 \cdot 9H_2O$, $MgSO_4 \cdot 7H_2O$, and $CaCl_2 \cdot 2H_2O$. All the chemicals were purchased from Sigma Chemical Co or EM Science. Fe(II) solution was prepared from ferrous ammonium sulfate (EM Science). Opaque polypropylene (brown) bottles were obtained from Cole-Parmer, USA and the polypropylene bottles (PP) and transparent poly(ethylene terephthalate) (PET) bottles were purchased from Fisher Scientific, USA. Philips black lights (20 W, 61 cm, PE 20 T12/BL) were used for UV source in the UV-A exposure experiments. A digital ultraviolet radiometer (Solarmeter, Model SM 5.0 UVA+B, USA) was used to measure the intensity of the UV radiation. A combined platinum redox electrode and a combined pH electrode (both from Fisher Scientific Co., USA) were used in this work.

INSTRUMENTATION

Most of the laboratory experiments were carried out flow injection hydride-generation atomic absorption spectroscopy (FI-HG-AAS) using a Perkin-Elmer (Model Zeeman 5000) atomic absorption spectrometer (AAS) equipped with an electrodeless discharge lamp (EDL) operated at 8W from an external power supply. The AAS was coupled with a Perkin-Elmer Flow Injection (FIAS-100) unit for hydride generation for the determinations of As(III) and As(Tot). Detail descriptions of the instrumental conditions are given in Table 2.1. Field samples for As(Tot) and speciated As(III) were analyzed by both FI-HG-AAS and ICP-MS (Sciex Elan 6000, Perkin Elmer, Norwalk, CT) equipped with a Gem-Tip cross-flow nebulizer, a four-channel peristaltic pump (Gibson, Model Minipuls III) and an Autosampler (As-90, Perkin Elmer, Norwalk, CT). Fe was measured by ICP-MS. Details of the instrumental set up are given in Table 2.2.

Table 2.1
Experimental conditions for the determination of As(III) and As(Tot) by FI-HG-AAS

Parameters	Perkin-Elmer (Zeeman 5000)
Lamp	EDL, powered by 8 (W)
Wavelength	193.7 (nm)
Slit	0.7 nm(low)
Sample volume	500 (μL)
HCl concentration	0.02 (M)
Citric/citrate buffer (pH 5.0)	2 (M)
HCl/citrate buffer flow rate	8 (ml/min)
$NaBH_4$ concentration (for As(Tot))	0.4% in 0.2% NaOH
$NaBH_4$ concentration [for As(III)]	0.2% in 0.05% NaOH
$NaBH_4$ flow rate	5 (ml/min)
Carrier gas	Argon
Carrier gas flow rate	60-70 (ml/min)
Quartz cell temp	Electrically heated, ~900 (°C)

Table 2.2
Parameters for ICP-MS operation

ICP system	
RF power	1300 W
Sampler and skimmer	Nickel
Nebulizer	Cross-flow
Spray chamber	Double-pass
Ar- plasma	15.0 (L/min)
Nebulizer-Ar	0.95–1.0 (L/min)
Analyte	^{75}As
For Correction	^{77}Se and ^{78}Se
Internal standard	^{127}Te

TOTAL ARSENIC AND ARSENIC(III) ANALYSIS

FI-HG-AAS

To determine As(Tot), samples were treated with L-cysteine in 2M HCl. The samples were kept for 15 minutes at room temperature for the reduction of As(V) to As(III) and then diluted to such a volume as to maintain the concentrations of L-cysteine and acid to 4 mg/mL and 0.02 M, respectively. The arsenic concentration was measured by FI-HG-AAS against As(III) standards prepared as samples. Reduced arsenic sample was injected into the stream of 0.02 M HCl (@ 8mL/min) solution by means of a rotary valve fitted with a 500μL sample loop. The injected sample together with carrier solution, met subsequently with a continuous stream of sodium tetrahydroborate (0.4% in 0.2 % NaOH, @ 5 mL/min). After mixing with sodium tetrahydroborate, the generated hydride (AsH_3) subsequently entered into the gas-liquid separator. Inside this apparatus a continuous flow of argon carrier gas (70-80 mL/min) carried the hydride to the quartz tube fitted on an electrically heated heater at 900 °C.

The peak height signals were measured at 193.7 nm. At least triplicate measurements were made for each sample and standard. For the determination of As(III) in the presence of As(V), the carrier HCl solution was replaced by 2 M citric/citrate buffer of pH 5.0. Arsine was generated using 0.2% sodium tetrahydroborate in 0.05 % NaOH. Under this condition As(V) does not interfere in the determination of As(III). The detection limits (3 times of standard deviation of the blank, n =10) for As(Tot) and As(III) were 0.04 μg/L and 0.096 μg/L, respectively. The relative standard deviations (RSD) of ten determinations of As(Tot) and As(III) were 1.86% and 3.34%.

ICP-MS

The ICP-MS was also used to measure the arsenic concentration in the preserved and field-speciated samples because, according to the PAC members, ICP-MS is more commonly used than FI-HG-AAS for arsenic analysis. The Perkin Elmer Sciex Elan 6000 ICP-MS was operated at 1300 W RF power using cross-flow nebulizer and 1 L/min nebulizer argon gas flow rate. All analyses were made in triplicate. Standards were prepared in 87 mM acetic acid and 1.34 mM EDTA, i.e., the matrix of the EDTA-HAc method.

Tellurium was used as an internal standard throughout the analysis. In ICP-MS, chloride interferes in the determination of arsenic due to formation of isobaric $^{40}Ar^{37}Cl^{+}$ molecular ions, which overlap at *m/z* with monoisotopic ^{75}As. The chloride interference was corrected using a standard mathematical correction equation (Amarasiriwardena et al. 1998). Intensities at *m/z* 77 and 82 were measured for the correction. The detection limit based on 3 times of the standard deviation of the blank (n =12) for As was 0.008μg/L. The relative standard deviation (RSD) of nine determinations of As (1 μg/L) in the presence of 1000 mg/L of chloride was within 2%. The recovery (n=6) for 5μg/L arsenic from the EDTA-HAc preserved NSF 2 challenge water was 103 ± 4%.

PREPARATION OF ARSENIC CONTAINING NSF 2 CHALLENGE WATER

For all laboratory experiments, NSF 2 Challenge Water (hereafter referred to as challenge water) whose composition is given in Table 2.3 was used because like the original NSF Challenge Water, it contains a median level of the typical anionic contaminants found in ground waters contaminated with arsenic and used for potable water supplies. However, the NSF 2 Challenge Water, which has lower calcium and magnesium concentrations, is more chemically stable than the original NSF Challenge Water, which began to precipitate calcium and magnesium after a few days (Wu MS thesis 2001). Stability was not an issue with the original NSF water because it is made up fresh daily. In contrast, this project required the challenge water to be stable for at least 28 days. Thus, the stability of the new NSF 2 Challenge Water was evaluated by the v 4.5 of MINEQL+ Chemical Equilibrium Modeling program (Schecher and McAvoy 2002). It showed no significant calcium or magnesium precipitation in the pH 6.5-8.5 range. During the experiments, the pH of the challenge water was adjusted to the desired pH in the 6.5 to 8.5 ranges by using dilute HCl and NaOH solutions.

Fe(II) in water gradually oxidizes to Fe(III) in the presence of dissolved oxygen, and subsequently precipitates as $Fe(OH)_3(s)$. Because arsenic tends to co-precipitate with the iron hydroxides it is important to keep the iron in Fe(II) state in the challenge water until preservative is added to the sample. The experimental set up is shown in Figure 2.1. Challenge water of pH ~8.3 was prepared in a brown glass bottle of 2.7 L capacity. The solution was sparged with N_2 for 1.5 h (In preliminary experiments dissolved oxygen was measured after sparging for 1.5 h in repeated experiments. The measured DO values were 56, 77, and 78 μg/L). As(III) and As(V) were spiked into the low-DO challenge water, and then freshly prepared Fe(II) solution was spiked under the sparging condition, so that the final concentration of Fe(II) in the challenge water was 3.0 mg/L. After 30-45 seconds, the first 500-mL (approx.) of the challenge water was drained through the sample collection outlet tube and then 500 mL of the challenge water was collected at maximum speed into a 500 mL volumetric flask containing the required amount of preservatives.

To study the effect of redox potential and pH on the distribution of arsenic species in the challenge water, laboratory experiments were designed using the brown bottle, sparger, platinum electrode etc. The experimental design was the same as before (Figure 2.1) with only addition of an Eh electrode. The redox potential of the challenge water was reduced to +120 ± 15 mV by sparging N_2 for 90 minutes. To reduce the redox potential further, a mixture of H_2 and N_2 (3:97) was sparged for some time to reach the Eh value –100 ± 10 mV. The flow rates were controlled by needle valves. If the Eh was far below the required Eh, compressed air and N_2 were sparged until it was set to the required Eh. To set the higher redox potential compressed air was sparged in to the challenge water after sparging with N_2. For our experiments, the Eh value of +200 ± 10 mV was maintained by controlling the flow of compressed air and N_2. The pH of the challenge water was studied in the range of 6.5-8.4. After attainment of a desired Eh value, Fe(II) was spiked into the challenge water under the sparging condition. Lowering of the Eh after addition of Fe(II) was accomplished by controlling the flow the gases. Then As(III) and As(V) were spiked, and after 30-45 seconds, the first 500 mL (approx.) of the challenge water was drained through the sample collection outlet tube and then 500 mL of the challenge water was collected at maximum speed into a 500-mL volumetric flask containing the required amount of preservative. The samples with preservatives were transferred into brown PP bottles and stored.

Samples were analyzed at predetermined time intervals to measure the concentrations of As(III) and total arsenic.

Table 2.3
Composition of NSF 2 Challenge Water

Cations	meq/L	mg/L	Anions	meq/L	mg/L
Na^+	3.604	82.9	HCO_3^-	2.0	122.0
Ca^{2+}	0.36	7.21	SO_4^{2-}	1.0	48.0
Mg^{2+}	0.26	3.16	Cl^-	0.36	12.8
			NO_3^--N	0.143	2.0
			F^-	0.053	1.0
			PO_4^{3-}-P	0.0013	0.04
			Silicate as SiO2	0.66	20.0
Total	4.224	93.27		4.224	205.8

Estimated TDS by evaporation = 287 mg/L

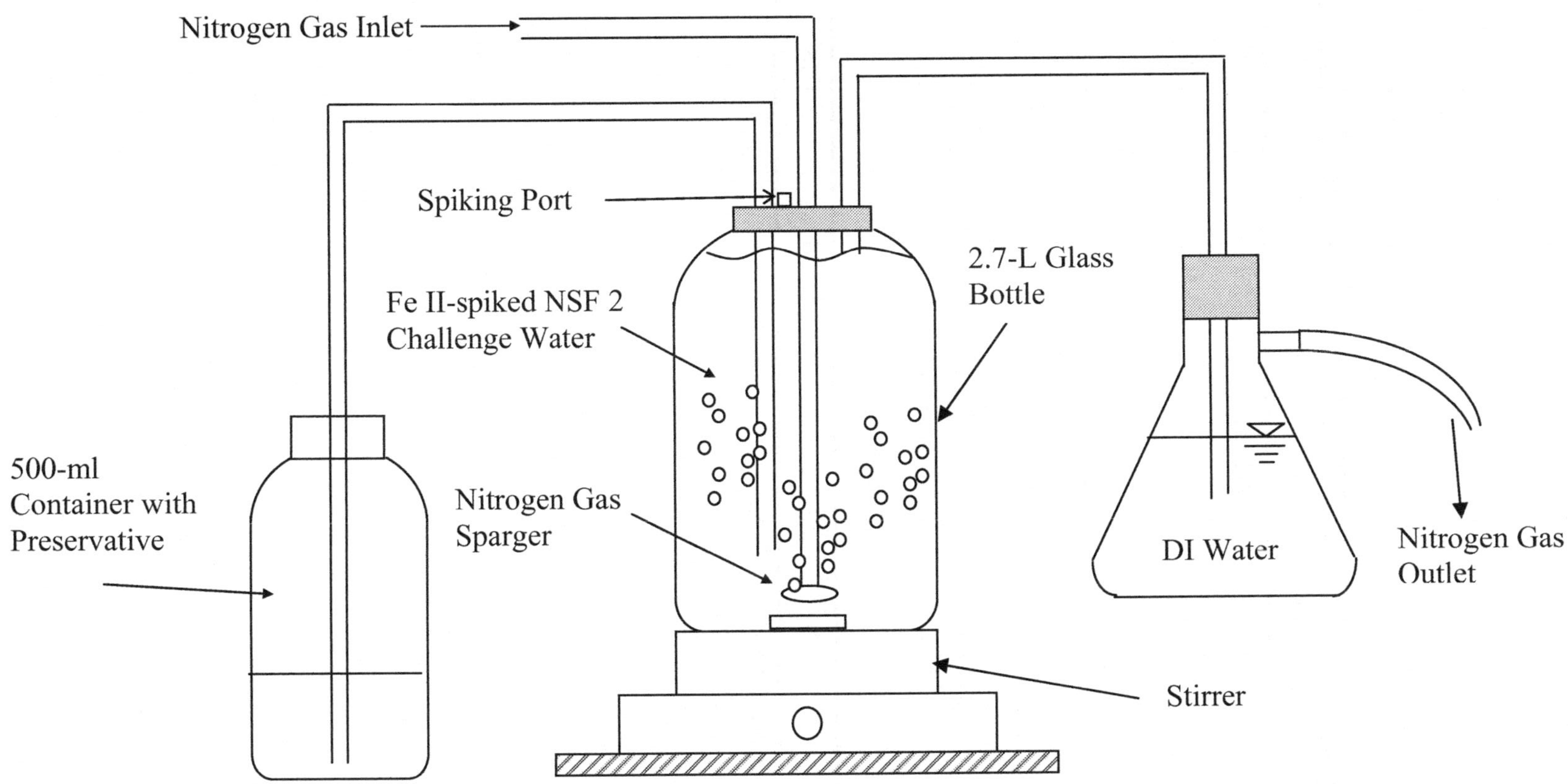

Figure 2.1. Apparatus for low DO challenge water with spiking port for Fe(II).

UV ILLUMINATION

To simulate solar irradiation in the laboratory, we used UV light (range of 300-400 nm) with similar spectral composition and intensity as sunlight in the spectral region where dissolved iron species absorb the main fraction of light. Up to six 1-L PET bottles (7 cm diameter) with 500 mL of solution were illuminated in the box with 4 Philips black-light tubes (20 W, 24", PE 20 T12/BL). See Figure 2.2 and 2.3. The peak maximum for lamp emission spectrum was 365 nm provided by the lamp manufacturer. The intensity of the incident UV radiation was 80 ± 10 W/m^2 measured by digital ultraviolet radiometer (Solarmeter, Model SM 5.0 UVA+B, USA) having maximum relative response at 375 nm.

During exposure, air circulation by a forced draft fan was maintained to control the temperature. Before exposure, As(III) concentrations in the samples with preservatives and without preservative (control) were measured, and then As(III) was measured after a specified time of exposure up to four hours. The temperature was monitored during exposure. The temperature of the waters in the PET bottles was initially 24 °C and increased to 29 °C during illumination in spite of the forced-air cooling. Duplicate samples were used for this experiment.

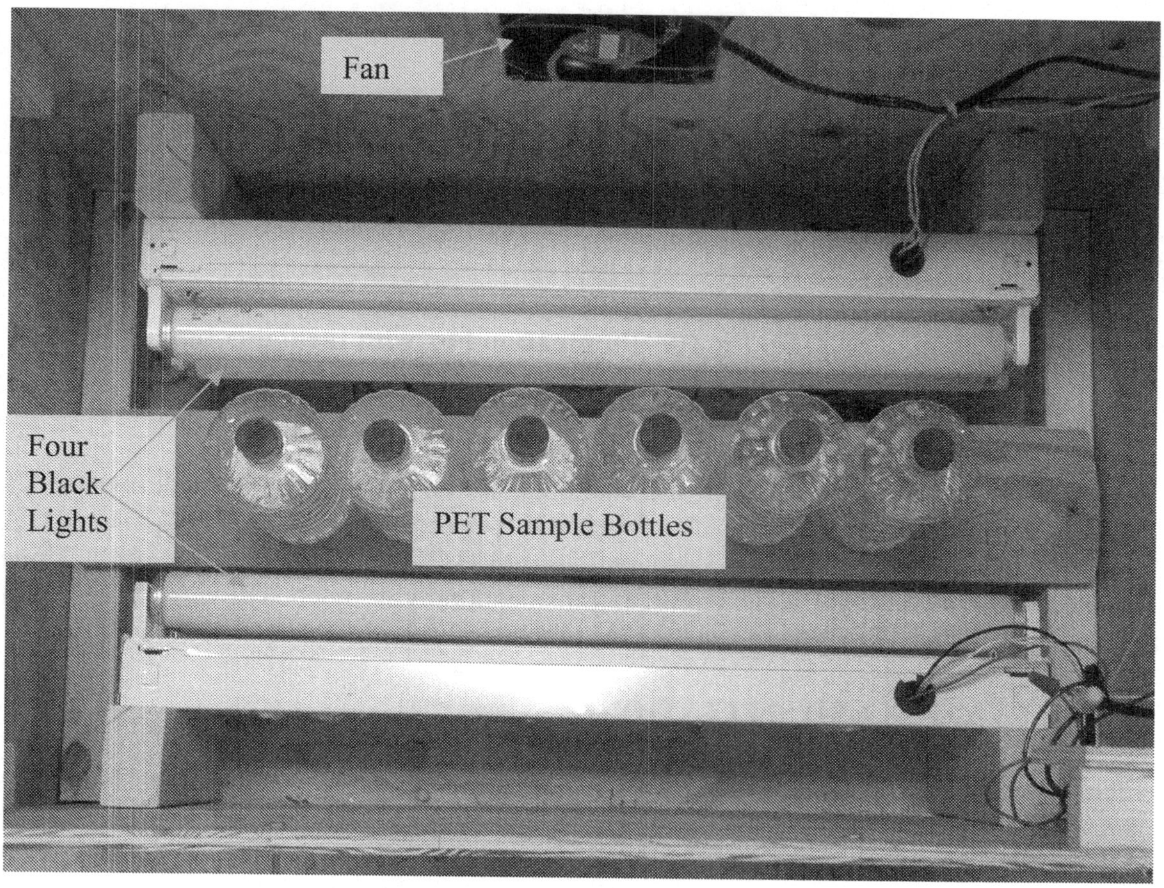

Figure 2.2. View looking down into the UV illumination box with the lights turned off.

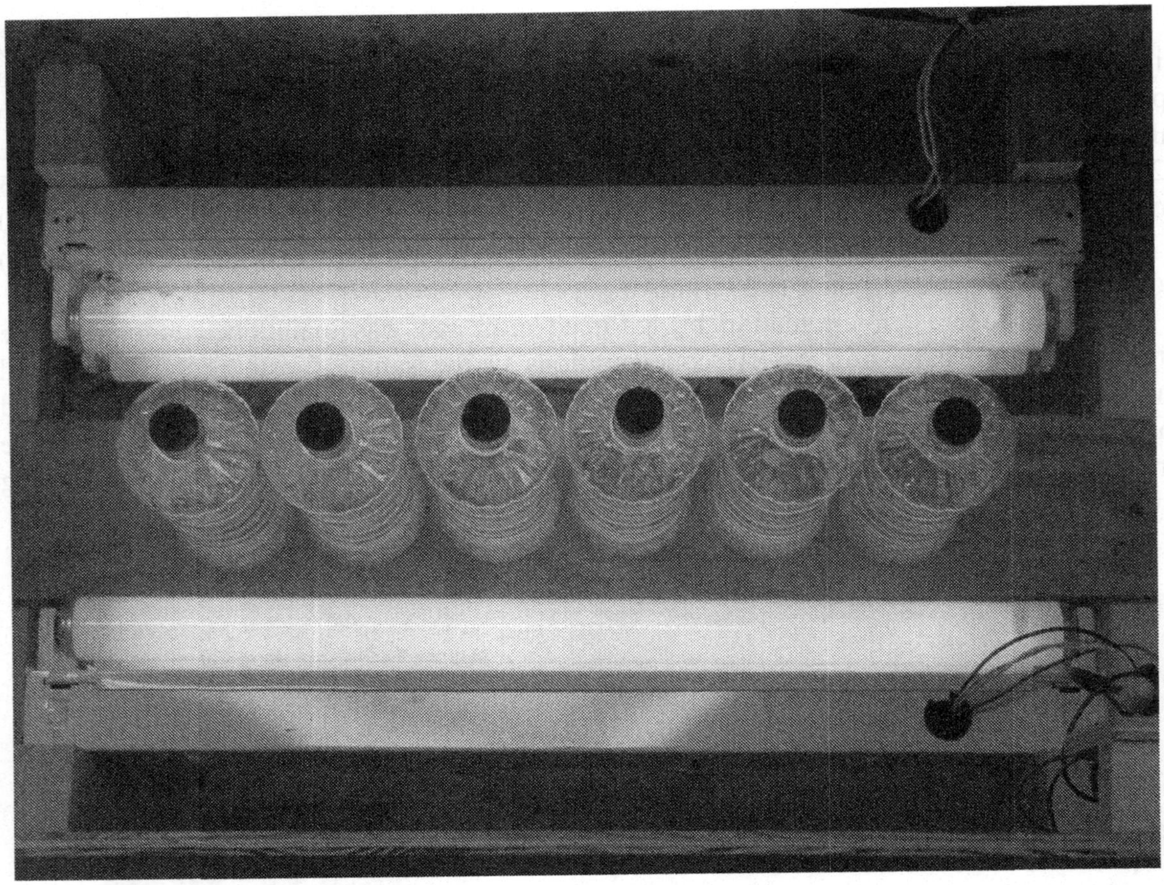

Figure 2.3. View looking down into the UV illumination box with the lights turned on.

PRESERVATIVES AND SAMPLE CONTAINERS

To preserve the arsenic species in the challenge water, attempts were made using 1.25 mM EDTA (no significant pH change), 10 mM H_2SO_4 (pH reduced to~1.95), 1.34 mM EDTA with 87 mM acetic acid (pH reduced to~3.15), and 20 mM H_3PO_4 (pH reduced to ~2.08). Three types of containers, brown polypropylene (BPP), translucent polypropylene (PP), and transparent polyethylene terephthalate (PET) were used for the laboratory experiments.

FIELD PRESERVATION AND SPECIATION

Field samples were collected directly into the 250-mL pre-cleaned brown (in few cases translucent) PP bottles, which contained 10.8 mL of 2M acetic acid and 3.35 mL of 0.1M EDTA. For field speciation, anion exchange resin column was used. Initially 10-mL Kontes Flex Glass minicolumn field with Cl-form resin was used. For speciation, 60-mL of the preserved sample was passed through the column; the first 15 mL of the effluent was discarded and the final 45 mL was collected in a cleaned plastic bottle. A detailed description of this column has been given in our previous AwwaRF report 2663 (2003).

The cost of this speciation column was ~ \$20.00 including resin and fittings. This was considered too high for routine speciation sampling at numerous sites, so the design was modified to reduce its cost.. New columns (11.0 cm long, 6.0 mm ID, and 1.2 cm OD) were made using polyvinyl tubing filled with chloride-form strong base anion resin (50x100 mesh Dowex 1X8-100, Supelco, PA). The resin bed was held in place between layers of glass wool at the column inlet and outlet. One end was fitted with a polypropylene slip female luer and the other end with a slip male luer fitting. The preserved sample was drawn into a 30-mL syringe, which was then mounted onto the top of speciation column through the female luer fitting. The first 8 mL of the sample passed through the resin column was discarded, and then 20 mL of the eluent was collected in a standard white polypropylene bottle. Uncharged arsenic species [primarily As(III)] remained in the eluent sample while charged As(V) species, remained in the resin column. As(III) in the eluent and As(Tot) in the preserved sample were measured in the laboratory. As(V) was calculated from the difference of As(Tot) and As(III). A detailed procedure for the field preservation and speciation method including drawing and photos is given in the Appendix A.

The validation of the EDTA-HAc arsenic species preservation technique was tested based on challenge water prepared in the laboratory under varying pH, redox, and UV light exposure conditions. The real test of the preservation technique must be based on actual ground water samples speciated in the field compared with laboratory speciation of the same samples in the lab up to 30 days later. In order to perform these fields vs. lab speciation tests, we received in-kind support from the University of Minnesota (approximately \$10,000), the Southern California Water Company (approximately \$1,600), and the City of Grand Forks North Dakota (\$1,750).

Additional University of Minnesota In-Kind Support

Melinda (Mindy) Erickson, a doctoral student in the Department of Civil and Environmental Engineering at the University of Minnesota, contacted us with a request to provide her with information on the arsenic III/V speciation testing procedure developed under the AwwaRF Grant Number 2663, "Field Separation Method for Arsenic Inorganic Species." We sent her a copy of the draft report, and with her advisor's approval, she adopted our ETDA-HAc ion-exchange speciation method. Her objective is to relate the speciation of arsenic to geological factors and water quality parameters in Minnesota drinking water wells. We agreed to provide her with arsenic analyses of field-speciated samples if she would send us preserved samples for our comparisons.

Using her grant funds, Mindy purchased the supplies necessary to make approximately 80 speciation kits according to the directions given in the draft AwwaRF report. Then she personally traveled to 48 well sites to take samples, preserve them, speciate the preserved samples, and send speciated and preserved samples to the University of Houston by overnight mail. Her field work has proven to be immensely beneficial to our project as it allowed us to test the EDTA-HAc preservation method on 65 actual samples from 48 drinking water wells, mostly contaminated with iron.

Additional Southern California Water Company In-Kind Support

Project Advisory Committee member David Kimbrough suggested we contact the Southern California Water Company for help in verifying the preservation method. We contacted them and received help through Dr. David Chang, Water Quality Manager, Region II. We prepared field-speciation and preservation kits, and sent them to Dr. Chang's staff who performed triplicate field speciation and preservation tests at four different well sites. Speciated and preserved samples were sent to the University of Houston by overnight mail. This additional opportunity to verify the preservation method was very helpful to the project, particularly because the California wells are so different compared with the Minnesota wells.

North Dakota Water Company In-Kind Support

As planned in our proposal to AwwaRF, we received help from Craig Lacher, Lab Supervisor of the City of Grand Forks. North Dakota Drinking Water Laboratory. We prepared field speciation and preservation kits and sent them to Craig, and he performed triplicate field speciation and preservation tests at three different well sites. Speciated and preserved samples were sent to the University of Houston by overnight mail.

CHAPTER 3
RESULT AND DISCUSSION

PRESERVATION OF ARSENIC SPECIES IN THE NSF 2 CHALLENGE WATER WITHOUT IRON

The stability of the arsenic species depends on several factors. Initially, both As(III)- and As(V)-containing challenge water (pH ~8.4) without iron was preserved with EDTA, EDTA-HAc, H_2SO_4, and H_3PO_4 at room temperature (22-24 °C) in brown polypropylene (BPP) bottles. As(III) was measured by hydride generation in the presence of citric/citrate buffer (pH 5.0) and As(V) was calculated from the difference between As(Tot) and As(III). The experimental results for EDTA, EDTA-HAc, H_2SO_4, and H_3PO_4 are shown in the Figures 3.1, 3.2, 3.3, and 3.4, respectively. Our experimental results indicate that As(III) and As(V) were stable for 168 h (7 d) when EDTA-HAc or H_2SO_4 or H_3PO_4 was added as a preservative. EDTA alone could not preserve the arsenic species for more than one day. After 24 h, As(III) was gradually oxidized to As(V). After 168 h (7d) more than 55% As(III) was oxidized to As(V). When no preservative (Figure 3.5) was added to the challenge water, As(III) was also stable up to 24 h and after 24 h it oxidized very rapidly. Within 48 h (2 d) more than 75% of As(III) was oxidized, and within 72 h, As(III) originally present in the challenge water, was converted to As(V). So, our experimental results indicate that EDTA alone was insufficient for preserving the distribution of arsenic species in the challenge water having no iron. This contradicts the observation of Bednar et al. (2002).

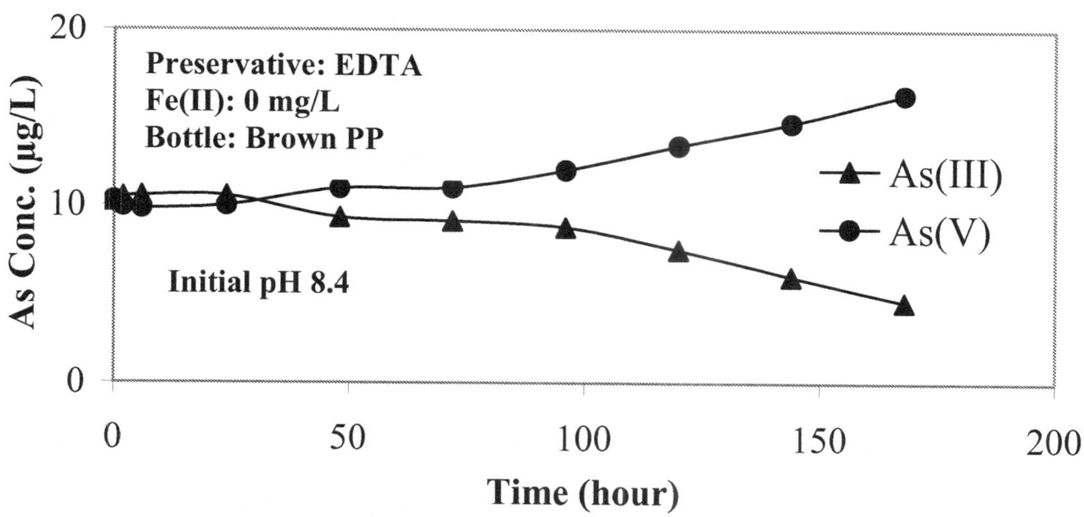

Figure 3.1 Stability of arsenic species with EDTA as a preservative in the NSF 2 challenge water (pH 8.4) having no Fe(II) in brown PP bottles at room temperature. The test solution initially contained 10 µg/L As(III) and 10 µg/L As(V).

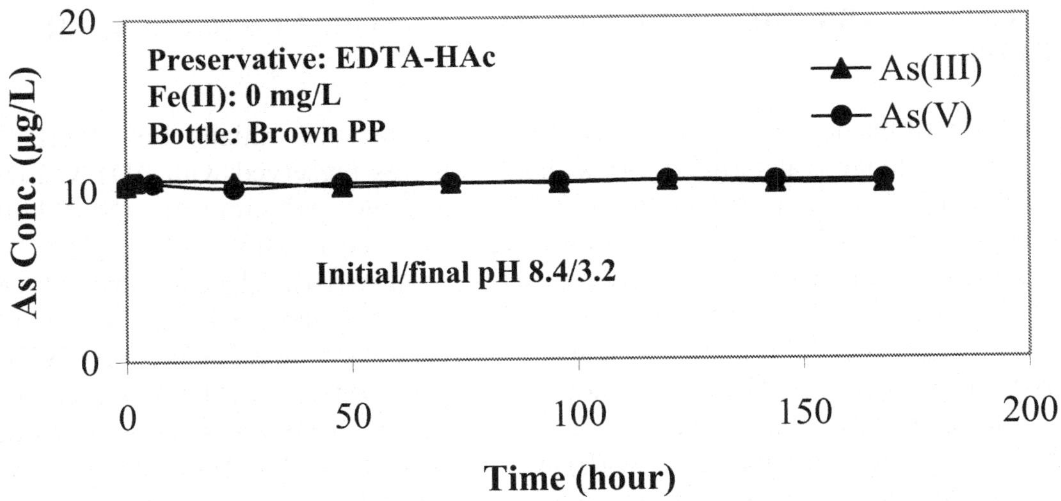

Figure 3.2 Stability of arsenic species with EDTA-HAc as a preservative in the NSF 2 challenge water (initial/final pH 8.4/3.2) having no Fe(II) in brown PP bottles at room temperature. The test solution initially contained 10 μg/L As(III) and 10 μg/L As(V).

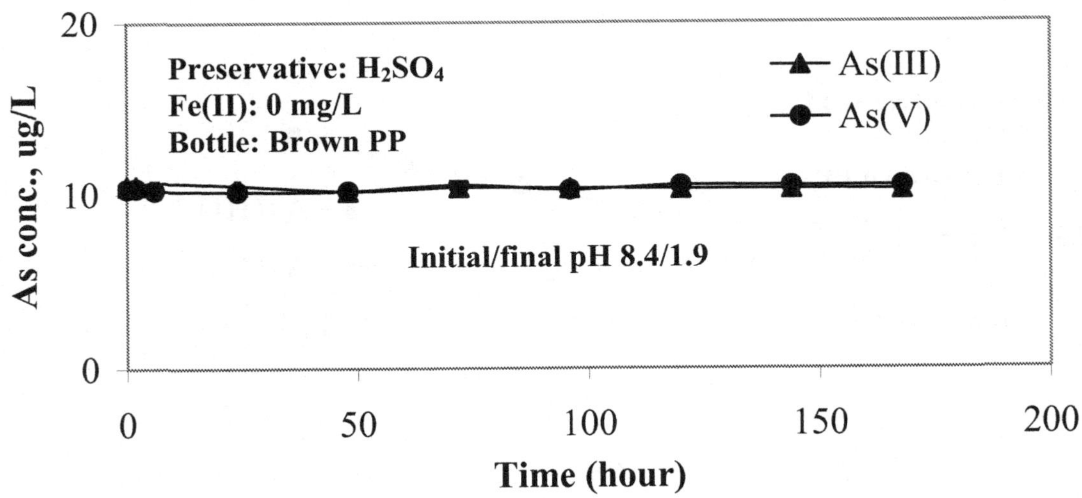

Figure 3.3 Stability of arsenic species with H_2SO_4 as a preservative in the NSF 2 challenge water (initial/final pH 8.4/1.9) having no Fe(II) in brown PP bottles at room temperature. The test solution initially contained 10 μg/L As(III) and 10 μg/L As(V).

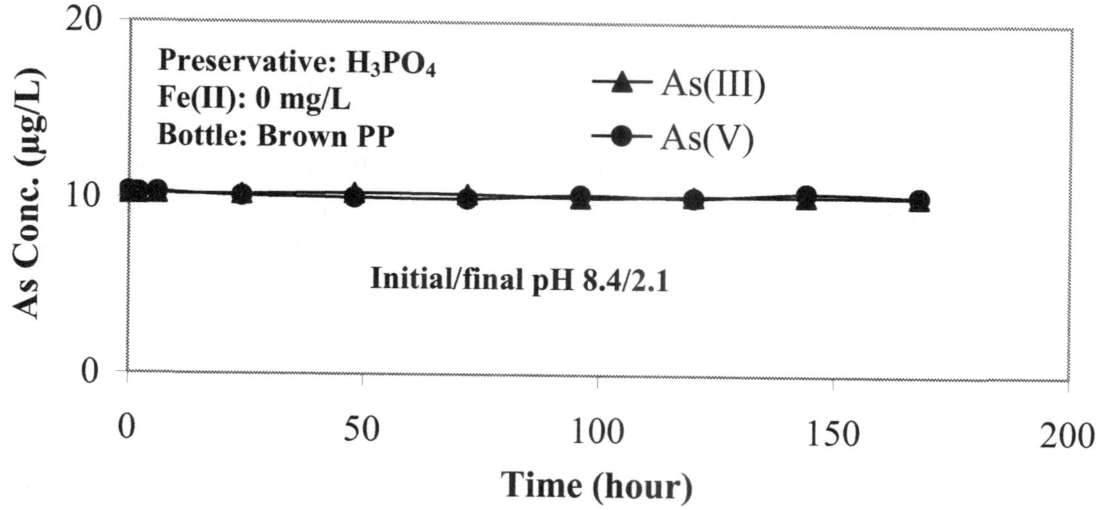

Figure 3.4 Stability of arsenic species with H_3PO_4 as a preservative in the NSF 2 challenge water (initial/final pH 8.4/2.1) having no Fe(II) in brown PP bottles at room temperature. The test solution initially contained 10 μg/L As(III) and 10 μg/L As(V).

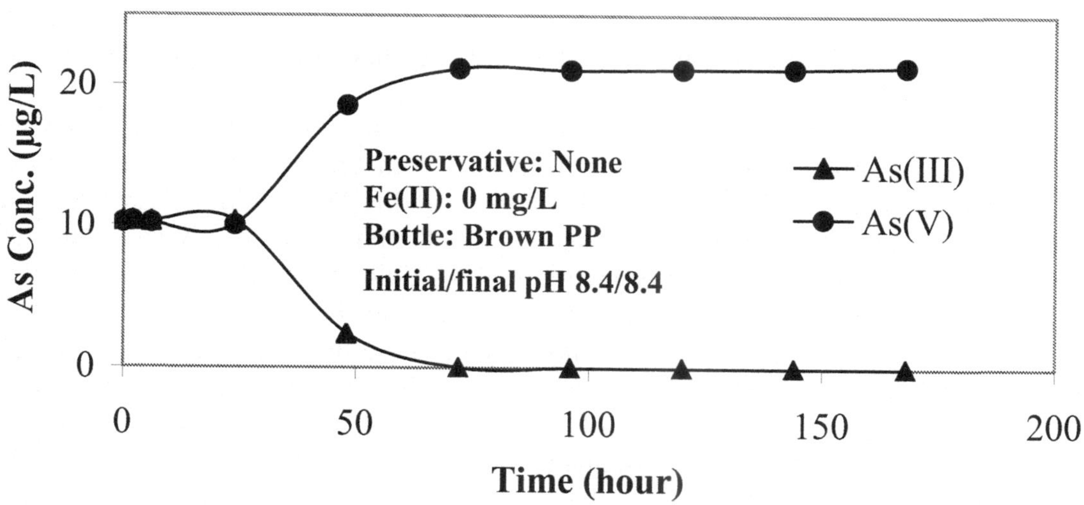

Figure 3.5 Stability of arsenic species without any preservative in the NSF 2 challenge water (initial/final pH 8.4/8.4) having no Fe(II) in brown PP bottles at room temperature. The test solution initially contained 10 μg/L As(III) and 10 μg/L As(V).

PRESERVATION OF ARSENIC SPECIES IN THE NSF 2 CHALLENGE WATER WITH IRON(II)

Preservation of Arsenic Species in Brown Polypropylene (BPP) Bottles

Experiments were carried out by spiking 3 mg/L Fe(II) into the challenge water (pH ~8.4) sparged with N_2. Arsenic containing challenge water with Fe(II) was preserved with EDTA-HAc, H_2SO_4, and H_3PO_4 at room temperature (22-24 °C) in BPP bottles. A control sample without any preservative was also collected in a BPP bottle. Sample with the preservative was colorless but sample without any preservative became brownish with time due to formation of the colloidal Fe(III)(hydr)oxides. Concentrations of As(III) were measured within two to three minutes.

It was observed that in the presence of Fe(II) at pH 8.4 without preservatives, As(III) was oxidized much faster (Figure 3.6) than in the absence of Fe(II) (Figure 3.5). Within three minutes of addition of Fe(II) in the challenge water, about 25-35% of As(III) was converted to As(V) and about 50-60% of As(III) was oxidized within 24 hours. A rapid decrease in the As(III) concentration was observed within the first few hours (approx. 5 h) and after that, the As(III) oxidation rate decreased. The provable cause of rapid oxidation of As(III) will be discussed later. According to Stumm and Morgan (1996), at groundwater pH values in the range of 7,2-8.0, Fe(II) was >90% oxidized within seconds to ten minutes and precipitated in the form of Fe(III)(hydr)oxides upon exposure to air. Others (Hug et al. 2001, Millero et al 1987, Lee and Choi 2002) found that, at a dissolved oxygen concentration of 2 mg/L, 90% Fe(II) was oxidized within 30 sec at pH 8.0 and within 1 hour at pH 7.0. It took about 100 hours at pH 6.0. So at the higher pH, the formation of Fe(III)(hydr)oxides is very fast, but the brown, metastable suspensions containing polynuclear Fe(III) complexes and poorly ordered Fe(III)(hydr) oxide phases do not precipitate for hours to days. Once Fe(III)(hydr)oxides are formed, the oxidation of As(III) to As(V) in the dark becomes very slow (Hug et al. 2001). Recently, Hug, et al. (2001) reported that the addition of 2 mg/L Fe(II) could oxidize As(III) at pH 7.8-8.0 almost completely within an hour.

Our experimental results also indicate that initially, the concentration of As(V) was increased due to oxidation of As(III) and then decreased with time due to co-precipitation of As(V) with Fe(III)(hydr)oxides. But, EDTA-HAc, H_2SO_4, and H_3PO_4 could preserve the arsenic species at room temperature for a long time (Figures 3.7, 3.8, and 3.9, respectively). The results show that EDTA-HAc (Figure 3.7) and H_2SO_4 (Figure 3.8) could preserve the arsenic species up 672 h (28 d) at room temperature without significant change in their distribution. But, H_3PO_4 (Figure 3.9) was not nearly as effective and could only preserve the distribution up to 168 h (7 d). After that As(III) converted slowly to As(V). According to Lytle and Snoeyink (2002) during period of mixing, Fe(II) is oxidized to Fe(III) and then Fe(III) reacts with phosphate to form ferric phosphate, which, under stagnant conditions, slowly dissolves to release orthophosphate and the ferric species (III) that are reduced to Fe(II). This process might influence the stability of the arsenic species in the Fe(II) containing challenge water preserved with H_3PO_4.

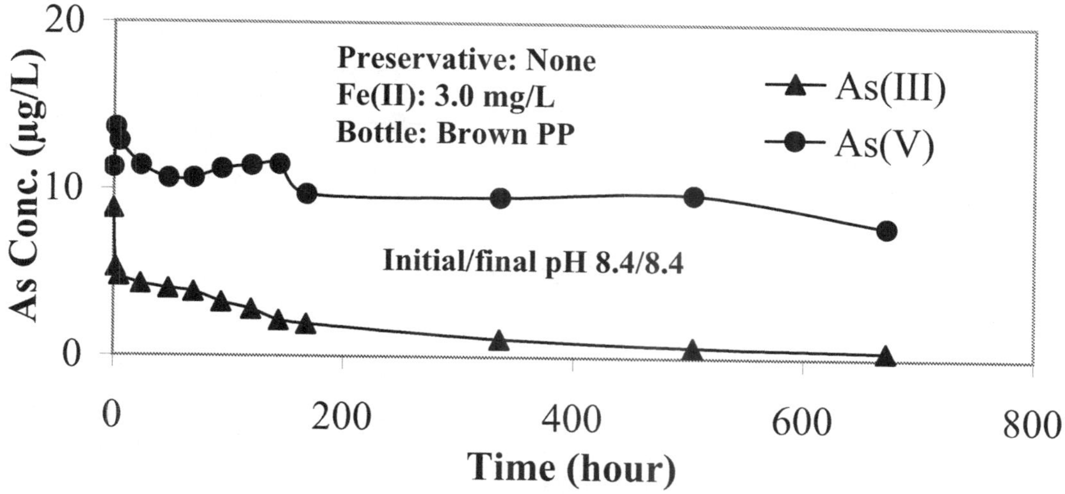

Figure 3.6 Stability of arsenic species without any preservative in the NSF 2 challenge water (initial/final pH 8.4) having 3 mg/L Fe(II) in brown PP bottles at room temperature. The test solution initially contained 10 µg/L As(III) and 10 µg/L As(V).

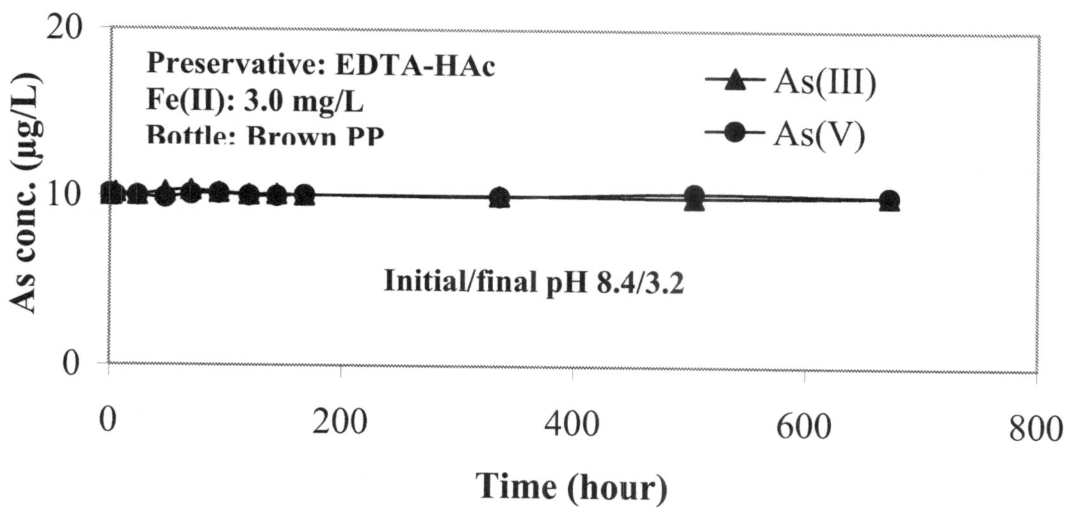

Figure 3.7 Stability of arsenic species with EDTA-HAc as a preservative in the NSF 2 challenge water (initial/final pH 8.4/3.2) having 3 mg/L Fe(II) in brown PP bottles at room temperature. The test solution initially contained 10 µg/L As(III) and As(V).

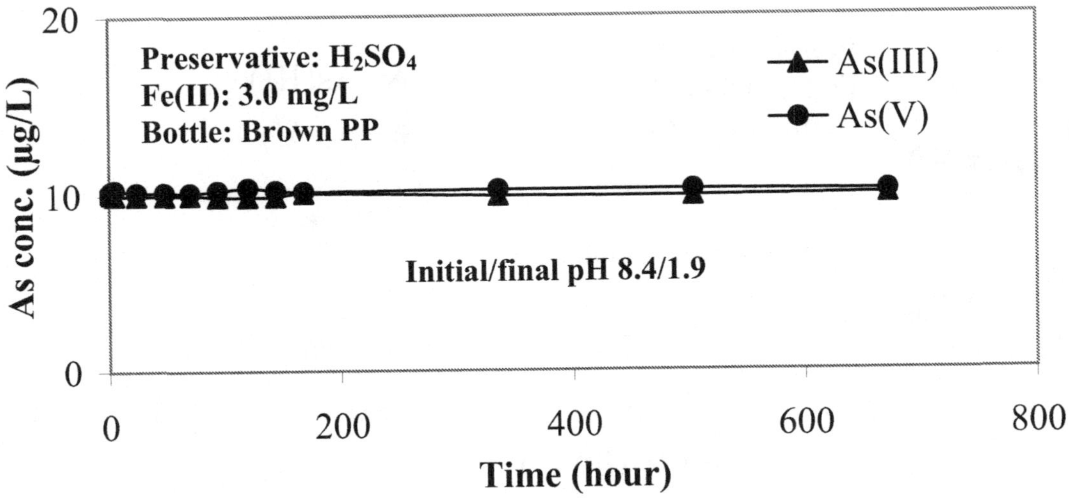

Figure 3.8 Stability of arsenic species with H_2SO_4 as a preservative in the NSF 2 challenge water (initial/final pH 8.4/1.9) having 3 mg/L Fe(II) in brown PP bottles at room temperature. The test solution initially contained 10 µg/L As(III) and 10 µg/L As(V).

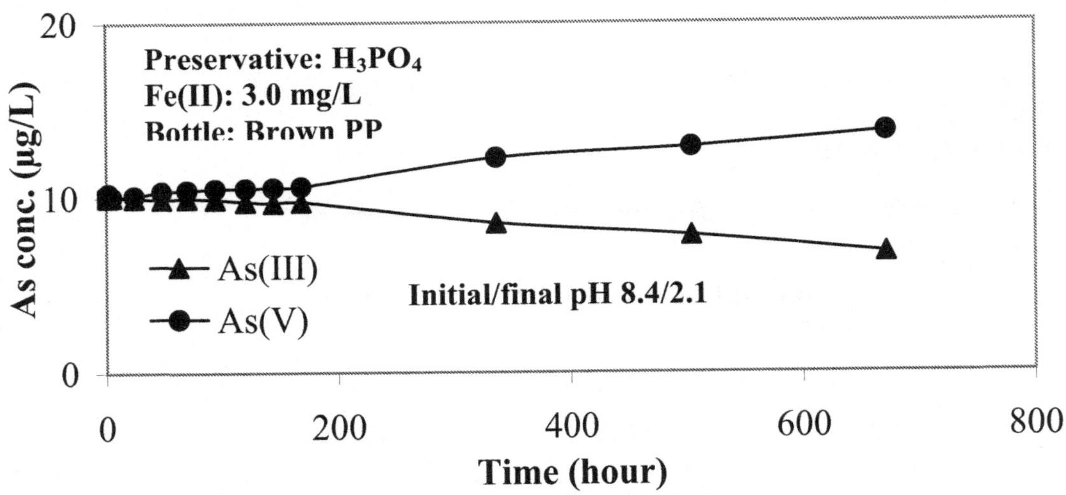

Figure 3.9 Stability of arsenic species with H_3PO_4 as a preservative in the NSF 2 challenge water (initial/final pH 8.4/2.1) having 3 mg/L Fe(II) in brown PP bottles at room temperature. The test solution initially contained 10 µg/L As(III) and 10 µg/L As(V).

Preservation of Arsenic Species in Crystal-Clear Transparent Polyethylene Terephthalate (PET) Bottles

The same experiment described above in brown polypropylene bottles was carried out preserving the arsenic-containing challenge water in crystal-clear transparent PET bottles exposed to typical laboratory fluorescent light. Samples with preservative were stored on top of a standard 36-inch-high laboratory bench under fluorescent lights mounted in ceiling fixtures six feet above the bench. The concentrations of As(III) and As(V) were monitored at predetermined time intervals up to 28 days. The results are shown in Figures 3.10, 3.11, and 3.12. The Figures show that H_3PO_4 (Figure 3.10) and H_2SO_4 (Figure 3.11) could preserve the arsenic species up to seven days, and EDTA-HAc (Figure 3.12) was an effective preservative for up to 21 days after which the distribution of the arsenic species shifted towards As(V).

Under the same experimental conditions, when the samples were preserved in brown PP bottles with EDTA-HAc or H_2SO_4, the distribution of arsenic species was constant up to 28 days. H_3PO_4 could preserve the species only up to 7 days in brown PP bottles or in clear PET bottles. From these experimental results, it was concluded that laboratory fluorescent light could play a role in the stability of arsenic species under typical storage conditions. To further examine the effect of UV light on the preserved samples, laboratory experiments were carried out in a wooden box fitted with a relatively intense UV-light source that simulated exposure of the samples to direct sunlight. The detail experimental set up has been discussed in Chapter 2.

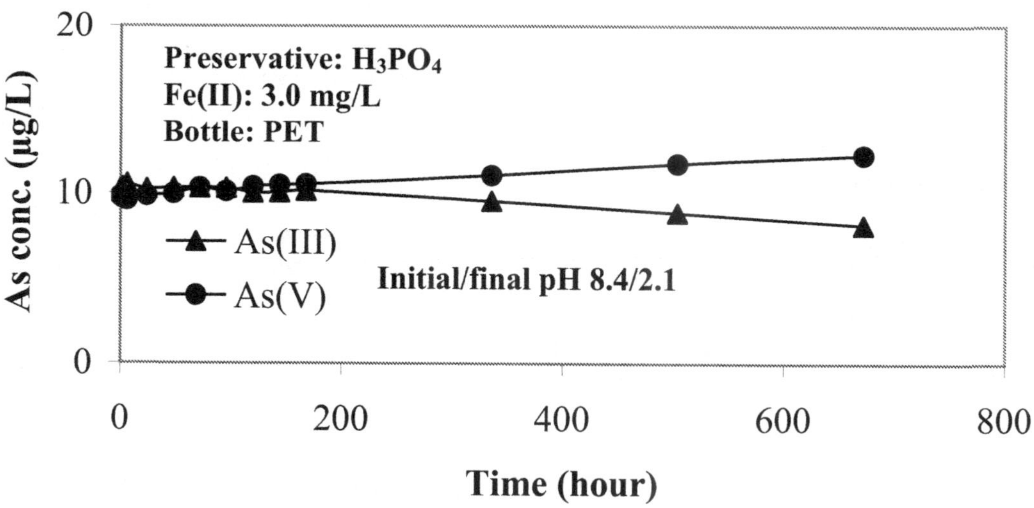

Figure 3.10 Stability of arsenic species with H_3PO_4 as a preservative in the NSF 2 challenge water (initial/final pH 8.4/2.1) having 3 mg/L Fe(II) in clear PET bottles at room temperature under laboratory light. The test solution initially contained 10 µg/L As(III) and 10 µg/L As(V).

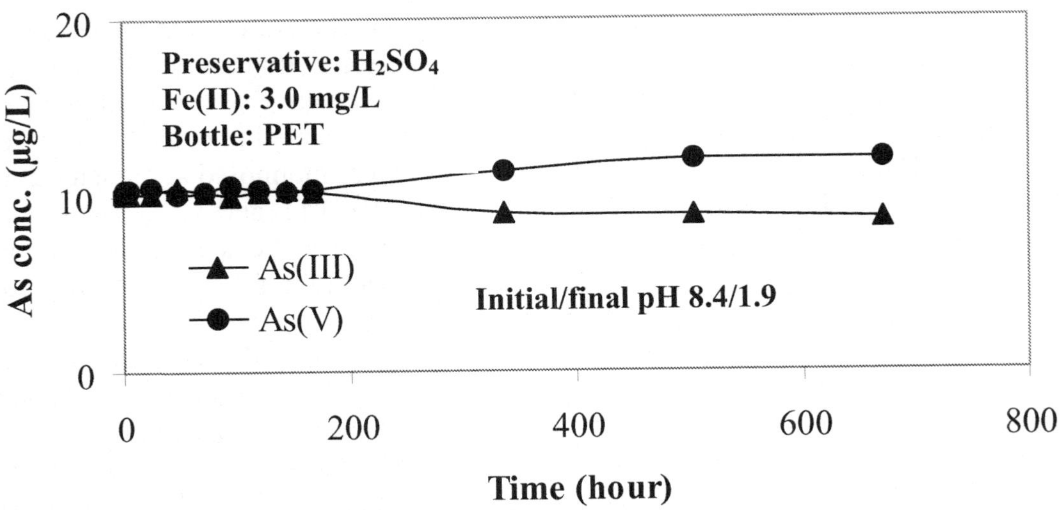

Figure 3.11 Stability of arsenic species with H_2SO_4 as a preservative in the NSF 2 challenge water (initial/final pH 8.4/1.9) having 3 mg/L Fe(II) in clear PET bottles at room temperature under laboratory light. The test solution initially contained 10 µg/L As(III) and 10 µg/L As(V).

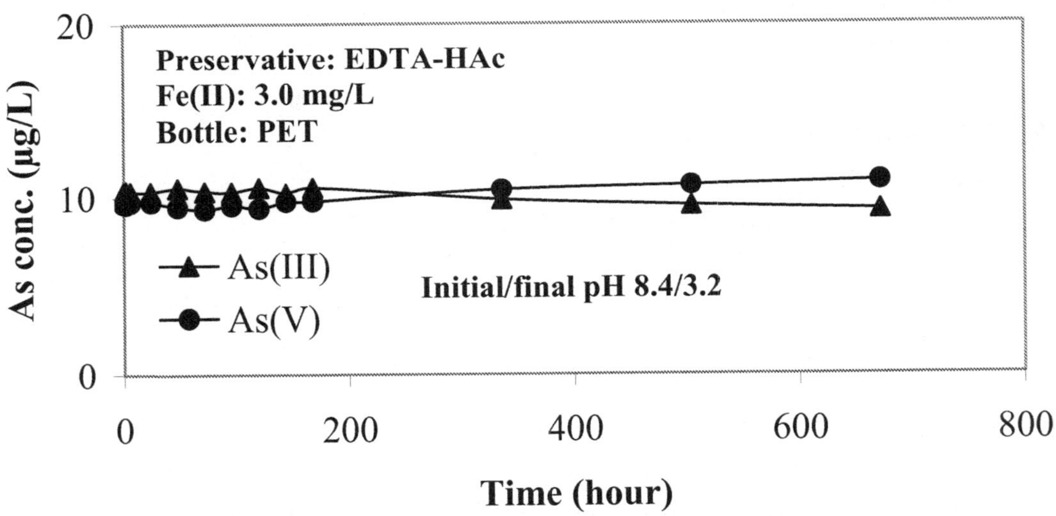

Figure 3.12 Preservation of arsenic species with EDTA-HAc as a preservative in the NSF 2 challenge water (initial/final pH 8.4/3.2) having 3 mg/L Fe(II) in clear PET bottles at room temperature under laboratory light. The test solution initially contained 10 µg/L As(III) and 10 µg/L As(V) each.

EFFECT OF SIMULATED SUNLIGHT UV ILLUMINATION ON ARSENIC (III) IN THE PRESENCE OF IRON(II)

One author's field-experience in West Bengal-India and Bangladesh where the light intensity of solar light between 300–400 nm is 80 ± 20 W/m^2 indicates that in most of the cases, samples were exposed to the sunlight during collection and transportation by walking from one field site to another. To study the effectiveness of these preservatives under simulated sunlight UV illumination with similar spectral composition and intensity, experiments were carried out in a UV-light box in the laboratory. The percentage As(III) oxidation resulting from with UV exposure in the laboratory using 500 mL challenge water (pH 8.4) with 3 mg/L Fe(II) in transparent PET bottles in the presence of three preservatives and with no preservative is shown in Figure 3.13. Before illumination and within three minutes after addition of Fe(II), As(III) concentrations were measured and then the samples were exposed to 300-400 nm UV light at an intensity of 80 ± 10 W/m^2 for designated times. The initial temperature of the sample was 24 °C. As(III) and temperature was measured after each predetermined time interval.

The initial measured As(III) concentration without any preservative was about 32% lower than the actual As(III) spiked into the challenge water, because at pH 8.4, the oxidation of Fe(II) to Fe(III) is very fast. The rapid oxidation of Fe(II) facilitates the As(III) oxidation as mentioned above. The chemical equations and rate constants from the literature that support the suggested mechanisms for the oxidation of As(III) to As(V) are shown in Table 3.1. Possible oxidants for As(III) oxidation (reactions R11-R17) are generated during oxidation of Fe(II) by oxygen (R1) and by other oxidants (R2, R3, R6, and R10). The major role for the oxidation of As(III) by the addition of Fe(II) may be due to the formation of the intermediate superoxides ($O_2^{\bullet -}$) by the reaction of Fe(II) and O_2 (Hug et al 2001, Lytle and Snoeyink 2002). The As(III) oxidation proceeds to As(V) through the intermediate As(IV) state. Once the intermediated As(IV) species are produced, they are rapidly converted to As(V) by dissolved O_2 (R15). On the other hand, in the presence of preservatives, As(III) was stable for a long period of time in the dark. Because of the formation of the stable complexes by the reaction of EDTA, H_2SO_4, and H_3 PO_4 with iron at lower pH (pH 1.78 –3.17) and at lower pH, Fe(II) is more resistant to oxidation.

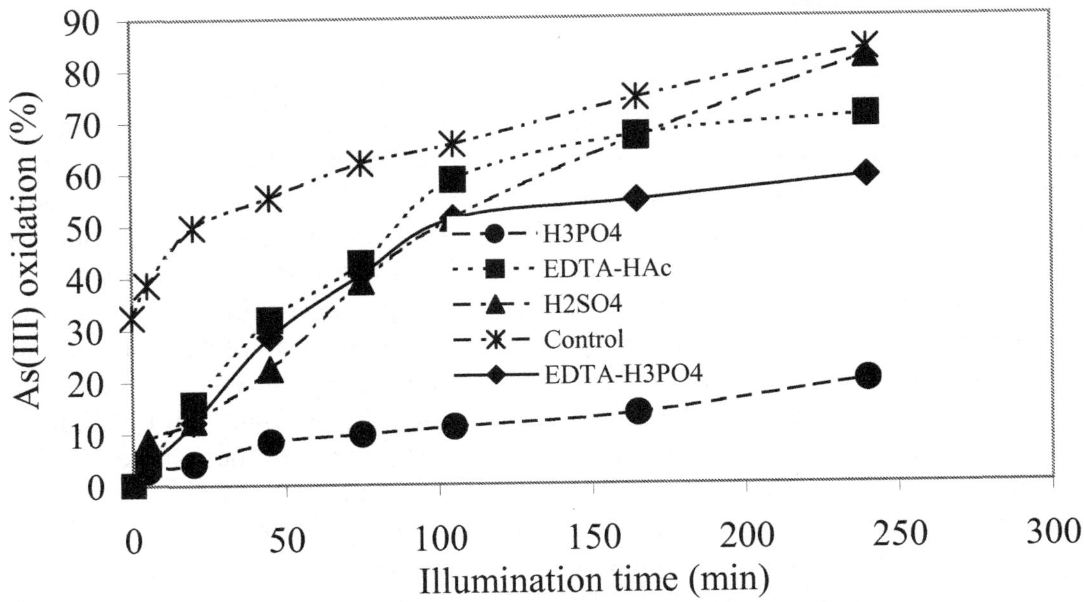

Figure 3.13. Percentage of As(III) oxidized by sunlight UV exposure with time in the challenge water (pH 8.4) having 3 mg/L Fe(II) preserved with EDTA-HAc, H_2SO_4, EDTA-H_3PO_4, and H_3PO_4 in clear PET bottles. The test solution initially contained As(III) and 10 μg/L As(V).

Figure 3.13 shows that about 50% of the spiked As(III) (measured before UV exposure) in the control sample [pH 8.4 and Fe(II) 3.0 mg/L] was oxidized within 20 min of UV exposure. Oxidation of As(III) was much faster at the beginning of the exposure. About 40% of As(III) was oxidized to As(V) within 5 min of exposure, and then the oxidation rate was comparatively slow. On the other hand, in the absence of UV illumination, it took about 24 h for approximately 50% oxidation of As(III) under the same experimental conditions (Figure 3.6). The rapid oxidation of As(III) under UV illumination is due to photolysis of $Fe(OH)^{2+}$(R18) and production of Fe(II), which is re-oxidized back to Fe(III) by oxygen and other oxidants. Fe(III) re-complexes (R7) and the cycle starts again. Figure 3.13 also shows that EDTA-HAc, H_2SO_4, and H_3PO_4 could not restrict the oxidation of As(III) in the presence of UV illumination.

The oxidation rates of As(III) in the presence of EDTA-HAc and H_2SO_4 were much higher than in the presence of H_3PO_4. One more experiment was carried out under the same experimental conditions in the presence of a mixture of EDTA-H_3PO_4 (1.34 mM EDTA and 20 mM H_3PO_4). From this result it can be concluded that EDTA is playing an important role in the oxidation of As(III) under the UV exposure. Within 20 min of exposure, about 16%, 9%, 12% and 4% of the As(III) was oxidized in the presence of EDTA-HAc, EDTA-H_3PO_4, H_2SO_4, and H_3PO_4 respectively. After 2 h, oxidation of As(III) in the presence of EDTA-HAc and EDTA-H_3PO_4 was not significant, but in the presence of H_2SO_4, and H_3PO_4 the oxidation of As(III) continued at about the same rate. During this period, the temperature increased approx. 2-3 °C. So, both photochemical and thermal oxidation may be responsible for the continued oxidation of As(III).

Again, the published list of iron and As(III) oxidation reactions in Table 3.1 is given as support for some possible reactions. EDTA forms a soluble complex with Fe(III), and the dissolved Fe(III)EDTA complex goes through photolysis (R19) under the UV illumination to reportedly produce EDTA radicals. These react readily with O_2, resulting ultimately in the formation of CH_2O, ethylenediaminetriacetate (ED3A), and $O_2\bullet^-$ (R20) (Karametaxas et al. 1995). Again, EDTA radical reacts with Fe(III)EDTA giving rise to another Fe(II) EDTA. The oxidation of As(III) is enhanced by the formation of $O_2\bullet^-$ due to photo degradation of Fe(III)EDTA complexes.

Table 3.1.
Relevant reactions with rate constants involved in arsenic oxidation process

#	**Iron-related reactions**	$k_n(M^{-1}S^{-1})$ or $k(M^{-1})$	Reference
R1	$Fe(II) + O_2 \rightarrow Fe(III)OH^{2+} + O_2^{\bullet-}$	60 (pH 7.5)	(Millero et al 1987)
R2	$Fe(II) + O_2^{\bullet-} + H^+ \rightarrow Fe(III) + H_2O_2$	1x 10^7	(Millero and Sotolongo 1989)
R3	$Fe(II) + H_2O_2 \rightarrow Fe(III) + \bullet OH + OH^-$	1.1x10^6	(Walling 1975)
R4	$Fe(III) + H_2O_2 \rightarrow Fe(II) + HO_2\bullet$	0.01 ~0.02	(Stuglik and Zagorski 1975)
R5	$Fe(II) + HO_2\bullet \rightarrow O_2 + Fe(II)$		
R6	$Fe(II) + \bullet OH \rightarrow Fe(III) + OH^-$	3.2x10^8	(Smith and Martell 1979)
R7	$Fe(III) \leftrightarrow Fe(III)OH^{2+} + H^+$	$pK_a = 2.3$	(Huie et al. 1991)
R8	$Fe(III)OH^{2+} \leftrightarrow Fe(III)(OH)_2^+ + H^+$	$pK_a = 3.6$	(Huie et al. 1991)
R9	$Fe(III)OH_2^+ \leftrightarrow Fe(III)(OH)_3^+ + H^+$	$pK_a = 7.9$	(Huie et al. 1991)
R10	$Fe(II) + CO_3\bullet^- \rightarrow Fe(III) + HCO_3^-$	3.6 x 10^8	(Buxton et al. 1988)
Arsenic Oxidation Reactions			
R11	$As(III) + O_2^{\bullet-} \rightarrow As(IV) + H_2O_2$	3.0 x 10^6	(Hug et al 2001)
R12	$As(III) + \bullet OH \rightarrow As(IV) + OH^-$	1.0 x 10^9 (pH 1-3)	(Pittine et al. 1999)
R13	$As(III) + H_2O_2 \rightarrow As(V) + 2\ OH^-$	15.3 (pH 10)	(Klaning et al. 1989)
R14	$As(III) + Fe(III) \rightarrow As(IV) + Fe(II)$		
R15	$As(IV) + O_2 \rightarrow As(V) + O_2^{\bullet-}$	1.1x 10^9	(Frank and Rau 1990)
R16	$As(IV) + Fe(III) \rightarrow As(V) + Fe(II)$		
R17	$As(III) + CO_3\bullet^- \rightarrow As(IV) + HCO_3^-$	1.1 x 10^8	(Frank and Rau 1990)
Photolysis & its follow-up reactions			
R18	$Fe(III)OH^{2+} + h\nu \rightarrow Fe(II) + \bullet OH$	4.0x10^{-5} @360 nm	(Hug et al 2001)
R19	$Fe(III)EDTA + h\nu \rightarrow Fe(II) + EDTAr$	1.1x10^{-4}	(Karametaxas et al 1995)
R20	$EDTAr + O_2 \rightarrow CH_2O + ED3A + O_2^{\bullet-}$	1.0x10^7	(Bondietti et al. 1993)
R21	$EDTAr + Fe(III)EDTA \rightarrow EDTAp + Fe(II)EDTA$	1.0 x 10^5	(Lindsay 1979)
R22	$Fe(II)EDTA + O_x \rightarrow Fe(III)EDTA + O_x$	1.0x10^5	(Karametaxas et al 1995)
R23	$+ HO_2\bullet + O_2^{\bullet-} \rightarrow H_2O_2$	9.7x10^7	(Millero et al 1987)

EDTAr stands for the decarboxylated EDTA; EDTAp stands for the cationic, oxidized radicals; ED3A stands for ethylenediaminetriacetate; O_x stands for the sum of the concentration of O_2 and other reactive oxygen species.

In the presence of H_2SO_4, the initial As(III) oxidation rate is slightly lower than the EDTA, and this may be due to slower dissociation of the iron sulfates compared with the dissociation of iron –EDTA complexes. In the presence of H_3PO_4 the rate of oxidation of As(III) under the UV illumination is comparatively slow. This may be due to the formation of more stable iron phosphates, which are not easily photo degradable under the UV exposure. So, it could be concluded that the oxidation of As(III) to As(V) depends on the UV exposure, temperature, and matrixes and samples should be preserved in opaque PP bottles to eliminate the influence of UV light.

PRESERVATION OF ARSENIC SPECIES IN THE CHALLENGE WATER WITH DIFFERENT REDOX POTENTIAL (E_H) AND PH

Redox potential (Eh) and pH are the most important factors controlling arsenic speciation. Both redox status and pH affect the speciation and the stability of arsenic species. Under oxidizing conditions, $H_2AsO_4^-$ is dominant at low pH (less than about pH 6.9), while at higher pH, $HAsO_4^{2-}$ becomes dominant. $H_3AsO_4^0$ and AsO_4^{3-} may be present in extremely acidic and alkaline conditions, respectively. Under reducing conditions at pH less than about 9.2, the uncharged arsenic species $H_3AsO_3^0$ will be predominant.

The predicted effect of Eh and pH on the oxidation state of arsenic is summarized in Figure 3.14, which was constructed using critically evaluated thermodynamic data (Lindsay 1979; Dove and Rimstidt 1985; Daus et al. 2002). To make the diagram more useful, the redox couples [Fe(III)/Fe(II)] are also included. The proposed hypothesis of As(III) oxidation is that Fe(II) is oxidized by the dissolved oxygen and produces superoxides and other oxidants, which react with As(III) to form an intermediate As(IV). Finally, As(IV) is oxidized to As(V) by oxygen. To stabilize the arsenic species in their original state, the redox activity of the accompanying redox couples like Fe(III)/Fe(II) should be minimized to interrupt the assumed cycle, i.e., the conversion of Fe(II) to Fe(III) by the reaction of O_2 should be minimized. The formation of metal complexes can help to decrease the free-metal-ion concentration (in reduced and/or oxidized form) in the solution. Additionally, the complex formation is able to prevent the precipitation of ferric compounds, which adsorb arsenic III and V. Therefore, different compounds like, EDTA-HAc, H_2SO_4, and H_3PO_4 at moderately acidic pH (1.78–3.17) were used. Table 3.2 shows the final pH after addition of the preservatives into the challenge water with the initial pH 8.4 to 6.5. Experiments to determine the effectiveness of the three preservatives were carried out at three initial pH values (8.4, 7.5, and 6.5), and two levels (-100 and + 200 mV) of redox potential.

Preservation of Arsenic Species under Reducing Conditions

For the reducing condition, challenge waters with pH 8.4, 7.5, and 6.5 were prepared and the redox potential was adjusted by controlling the flow of H_2 and N_2 such that a potential of minus 100 ± 10 mV was attained. As(III) and As(V) were spiked into the challenge water after spiking it with 3 mg/L Fe(II). Immediately after spiking, samples for arsenic analysis were taken in brown PP bottles with and without preservative added. (The control sample at each Eh-pH condition had no preservative.) The initial Eh values of the samples with and without preservative were measured in the brown bottles within 30 minutes of taking the sample. The Eh values of the control sample (without preservative), and samples with H_2SO_4, H_3PO_4, and EDTA-HAc were in the range of 105-115 mV, 525-536 mV, 397-412 mV and 374-389 mV respectively. As(III) (determined by arsine generation in the presence of citric/citrate buffer at pH 5) and total arsenic (determined by arsine generation in the presence of HCl after the reduction of As(V) to As(III) by L-cysteine) were measured in each sample with time. Triplicate As(III) and As(Tot) determinations were made for each sample and averaged. The results are tabulated in Tables 3.3-3.6 and graphically presented in Figures 3.15-3.26 for pH in the range of

6.5 to 8.4. Within ± 10% deviation of the initial As(III) concentration was considered as an effective performance of the preservatives.

The performance of H_3PO_4 to stabilize the As(III) is shown in Table 3.3 and Figures 3.15, 3.19, and 3.23, which show that H_3PO_4 can preserve arsenic species up to 28 days, 21 days and 3 days for initial sample pHs of 8.4, 7.5 and 6.5, respectively. H_2SO_4 (Table 3.4 and Figures 3.16, 3.20, and 3.24) can preserve arsenic species at least up to 21 days without significant change. But the best preservation was accomplished by, EDTA-HAc (Table 3.5 and Figures 3.17, 3.21, and 3.25), which could preserve the arsenic species up to 28 days without any significant change.

In summary, H_3PO_4 and H_2SO_4 could preserve arsenic species up to 21 days at room temperature in samples having initial pH range 6.5 – 8.4 with low Eh (-100 ± 10 mV) except for pH 6.5. We could not find any proper reason for such deviation. However, EDTA-HAc was an effective preservative at all three pHs values for up to 28 days. According to the Eh-pH diagram (Figure 3.14) within the Eh range 375 to 525 mV with pH range 1.80 to 3.10, H_3AsO_3 is the most favorable arsenic species. So, under reducing condition, all the three preservatives can preserve arsenic species for three weeks or more.

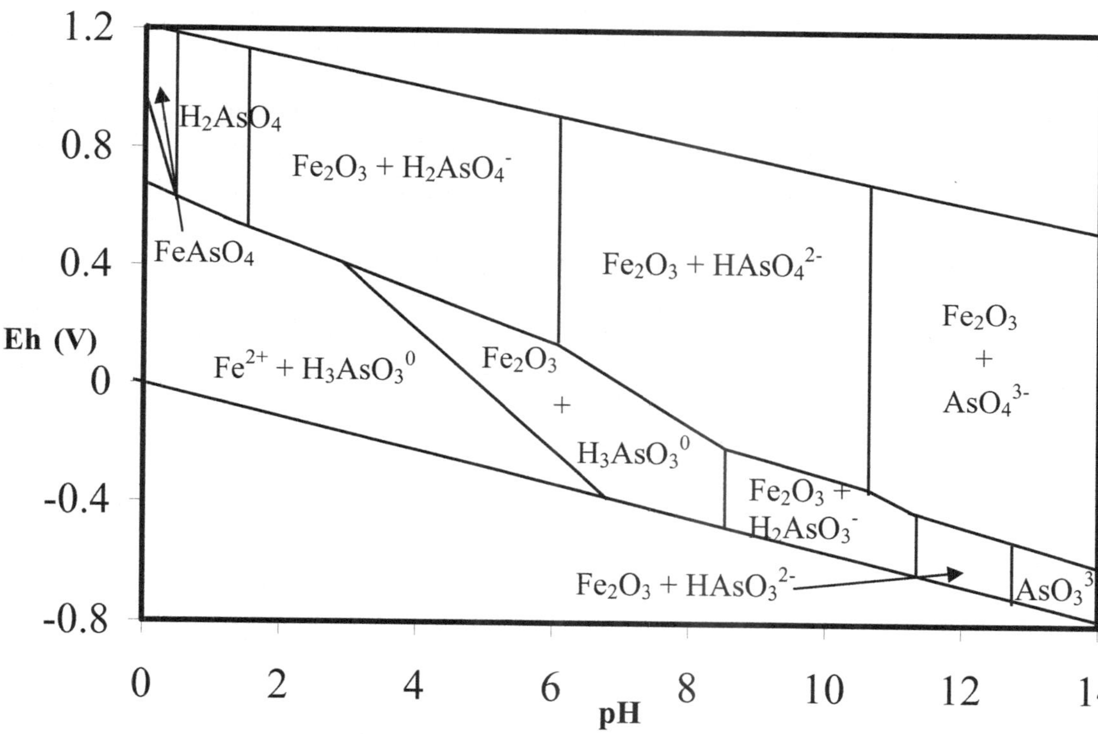

Figure 3.14. Eh-pH diagram for the system As-Fe-H_2O, $\sum As = \sum Fe = 10^{-6}$, modified after Daus et al. 2002

Table 3.2
Initial and final pH of the experimental waters

Initial pH	Preservative added	Final pH
8.4	H_3PO_4	2.07 ± 0.06
8.4	H_2SO_4	1.88 ± 0.05
8.4	EDTA-HAc	3.17 ± 0.02
7.5	H_3PO_4	2.05 ± 0.03
7.5	H_2SO_4	1.82 ± 0.04
7.5	EDTA-HAc	3.13 ± 0.02
6.5	H_3PO_4	2.05 ± 0.06
6.5	H_2SO_4	1.78 ± 0.08
6.5	EDTA-HAc	3.06 ± 0.03

For control samples, (Table 3.6 and Figures 3.18, 3.22, and 3.26) initially the As(III) concentration was decreased, but after 5 h the As(III) concentration was increased and stable for a few days at pH 6.5 (Figure 3.26) and 8.4 (Figure 3.18) and then it started to decease. In case of pH 8.4, As(III) concentration decreased rapidly. At first, the samples were under reducing condition (-100 ± 10 mV) but during transfer of the samples into brown PP bottles, the Eh value increased to +100 ± 10 mV in contact with air. So, rapid oxidation occurred until equilibrium was achieved. According to the Eh-pH diagram (Figure 3.14), below the neutral pH, As(III) is more stable as $H_3AsO_3^0$ at Eh around +100 mV, where as, at higher pH, As(V) is more stable and present as $HAsO_4^{2-}$. Our experimental results show that at pH 6.5 (Figure 3.26) after equilibrium (after 24 h) As(III) was stable for a longer period of time than the pH 7.5 (Figure 3.22) and pH 8.5 (Figure 3.18) and then gradually oxidized to As(V) with time. However, it can be concluded that arsenic species are not stable in the pH range 6.5 to 8.4 under reduced redox potential. In all cases the As(Tot) concentration decreased with time and it was due to adsorption of arsenic species on Fe(III)(hydr)oxides precipitates.

Table 3.3
Performance of H_3PO_4 under reducing condition (-100 ± 10 mV) at different pHs.

Initial pH	Time, h	As(III) conc. μg/L (n= 3)	Deviation (%)
8.4	0.05	10.13	-
	2	10.06	0.66
	4	10.13	0.00
	24	10.43	2.99
	48	10.59	4.65
	72	10.19	0.62
	96	10.20	0.74
	120	9.83	2.98
	144	9.81	3.15
	168	9.60	5.21
	336	9.99	1.35
	504	9.499	6.21
	672	9.92	2.05
7.5	0.05	10.16	-
	4	10.05	0.99
	24	9.95	2.04
	48	9.61	5.39
	72	9.91	2.46
	96	9.91	2.39
	120	9.50	6.28
	144	9.53	6.12
	168	9.87	2.87
	336	9.54	6.05
	504	9.18	9.62
	672	8.91	12.27*
6.5	0.05	9.99	-
	2	9.74	2.55
	4	9.88	1.12
	24	9.58	4.14
	48	10.03	0.36
	72	9.05	9.42
	96	8.83	11.59*
	120	8.63	13.65*
	144	8.66	13.28*
	168	8.34	16.49*
	336	7.91	20.81*
	504	7.26	27.32*
	672	5.95	40.47*

*indicates that the preservative failed because the deviation exceeded ± 10%.

Table 3.4
Performance of H_2SO_4 under reducing condition (-100 ± 10 mV) at different pHs.

Initial pH	Time, h	As(III) conc. µg/L (n= 3)	Deviation (%)
	0.05	9.97	-
	2	9.98	0.01
	4	9.89	0.91
	24	10.24	2.54
	48	10.35	3.64
	72	10.23	2.47
8.4	96	10.18	2.03
	120	10.25	2.75
	144	10.05	0.81
	168	9.82	1.62
	336	9.98	0.04
	504	9.96	0.25
	672	10.40	4.18
	0.05	9.99	-
	4	9.90	0.82
	24	9.76	2.22
	48	9.58	4.03
	72	9.69	2.86
	96	9.70	2.85
7.5	120	9.65	3.36
	144	9.52	4.68
	168	9.88	1.08
	336	10.05	0.61
	504	9.33	6.59
	672	10.00	0.20
	0.05	9.43	-
	2	9.54	1.13
	4	9.58	1.53
	24	9.24	2.08
	48	9.16	2.95
	72	9.11	3.39
6.5	96	8.98	4.83
	120	8.80	6.67
	144	9.06	3.99
	168	9.08	3.75
	336	9.25	1.92
	504	8.92	5.42
	672	8.41	10.89*

*indicates that the preservative failed because the deviation exceeded ± 10%.

Table 3.5
Performance of EDTA-HAc under reducing condition (-100 ± 10 mV) at different pHs.

Initial pH	Time, h	As(III) conc. µg/L (n= 3)	Deviation (%)
	0.05	9.51	-
	2	9.53	0.13
	4	9.16	3.76
	24	9.57	0.55
	48	9.86	3.62
	72	10.03	5.40
8.4	96	9.97	4.81
	120	9.49	0.21
	144	9.31	2.11
	168	9.38	1.44
	336	9.72	2.14
	504	9.47	0.45
	672	9.67	1.64
	0.05	9.98	-
	4	9.85	1.33
	24	10.00	0.22
	48	9.98	0.06
	72	10.13	1.55
	96	9.80	1.75
7.5	120	9.94	0.41
	144	9.95	0.31
	168	9.97	0.11
	336	10.28	3.05
	504	10.07	0.89
	672	10.05	0.71
	0.05	9.58	-
	2	9.43	1.51
	4	9.43	1.53
	24	9.69	1.27
	48	9.64	0.73
	72	9.81	2.48
6.5	96	9.65	0.78
	120	9.99	4.42
	144	9.79	2.20
	168	9.77	1.99
	336	9.66	0.89
	504	9.85	2.86
	672	9.61	0.38

Table 3.6
Performance without preservative under reducing condition (-100 ± 10 mV) at different pHs.

Initial pH	Time, h	As(III) conc. μg/L (n= 3)	Deviation (%)
	0.05	6.69	-
	2	5.74	14.34*
	4	5.79	13.46*
	24	5.60	16.32*
	48	12.22	82.48*
	72	13.08	95.34*
8.4	96	12.69	89.48*
	120	7.02	4.81
	144	3.35	49.97*
	168	1.74	73.98*
	336	1.56	76.65*
	504	1.46	78.16*
	672	1.29	80.77*
	0.05	9.81	-
	4	5.85	40.37*
	24	5.84	40.43*
	48	9.64	1.68
	72	11.93	21.66*
	96	11.29	15.23*
	120	9.89	0.87
7.5	144	8.76	10.69*
	168	8.09	17.42*
	336	3.65	62.76*
	504	0.000	100*
	672	0.000	100*
	0.05	7.21	-
	2	5.34	25.98*
	4	5.24	27.34*
	24	4.65	35.53*
	48	7.58	5.07
	72	7.07	1.96
6.5	96	6.86	4.86
	120	6.70	7.05
	144	6.50	9.84
	168	6.61	8.36
	336	1.75	75.72*
	504	1.06	85.28*
	672	0.62	91.41*

*indicates that the preservative failed because the deviation exceeded ± 10%.

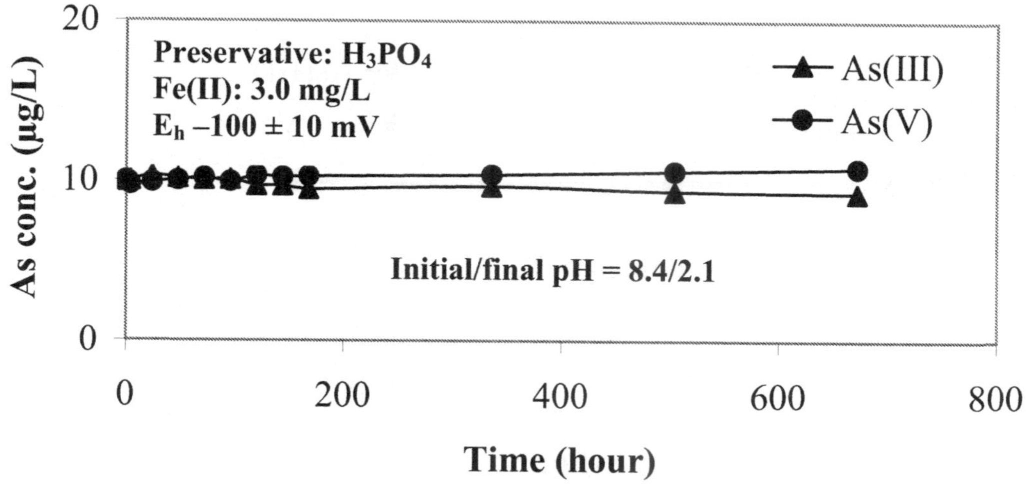

Figure 3.15 Stability of arsenic species in NSF 2 challenge water (initial/final pH 8.4/2.1) containing 3 mg/L Fe(II) with Eh –100 ± 10 mV using H_3PO_4 as a preservative in a brown PP bottle at room temperature. The test solution initially contained 10 μg/L As(III) and 10 μg/L As(V).

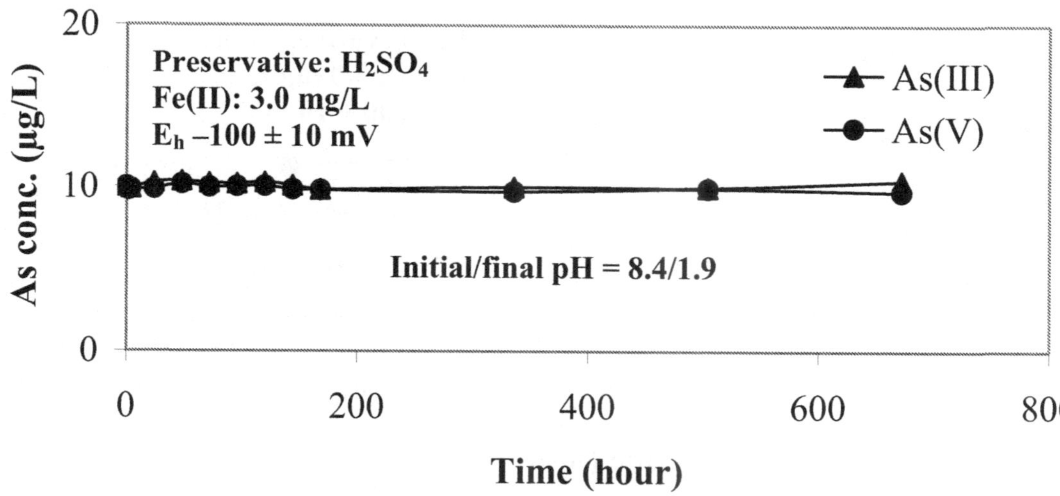

Figure 3.16 Stability of arsenic species in NSF 2 challenge water (initial/final pH 8.4/1.9) containing 3 mg/L Fe(II) with Eh –100 ± 10 mV using H_2SO_4 as a preservative in a brown PP bottle at room temperature. The test solution initially contained 10 μg/L As(III) and 10 μg/L As(V).

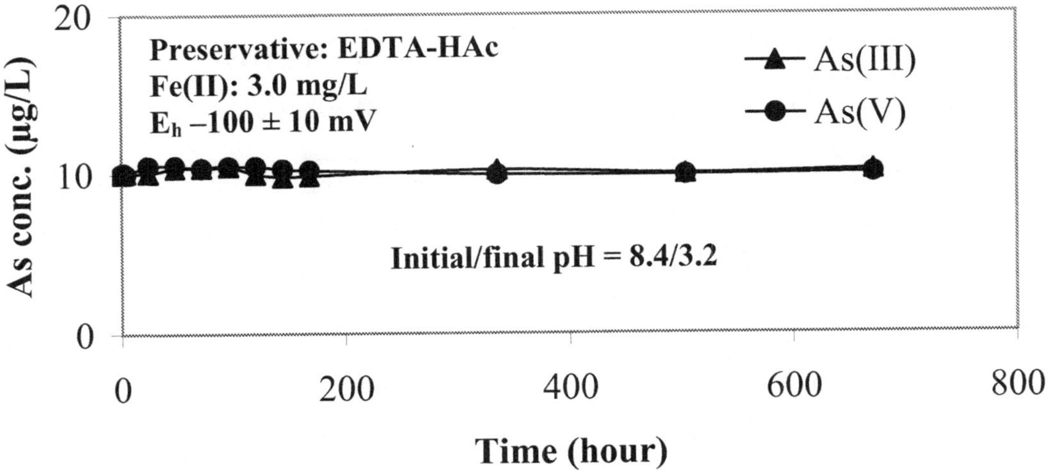

Figure 3.17 Stability of arsenic species in NSF 2 challenge water (initial/final pH 8.4/3.2) containing 3 mg/L Fe(II) with Eh –100 ± 10 mV using EDTA-HAc as a preservative in a brown PP bottle at room temperature. The test solution initially contained 10 µg/L As(III) and 10 µg/L As(V).

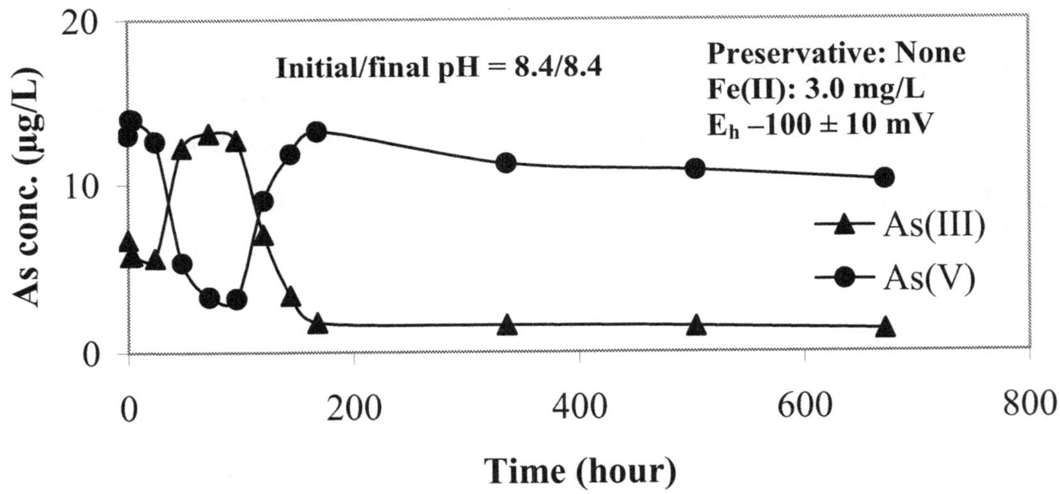

Figure 3.18 Stability of arsenic species in NSF 2 challenge water (initial/final pH 8.4/8.4) containing 3 mg/L Fe(II) with Eh –100 ± 10 mV using no preservative in a brown PP bottle at room temperature. The test solution initially contained As(III) and 10 µg/L As(V).

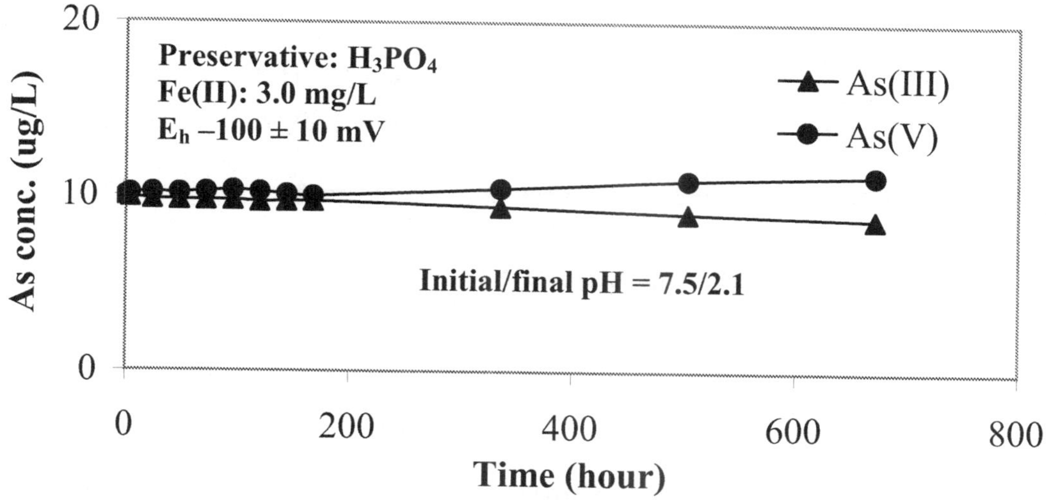

Figure 3.19 Stability of arsenic species in NSF 2 challenge water (initial/final pH 7.5/2.1) containing 3 mg/L Fe(II) with Eh −100 ± 10 mV using H_3PO_4 as a preservative in a brown PP bottle at room temperature. The test solution initially contained 10 µg/L As(III) and 10 µg/L As(V).

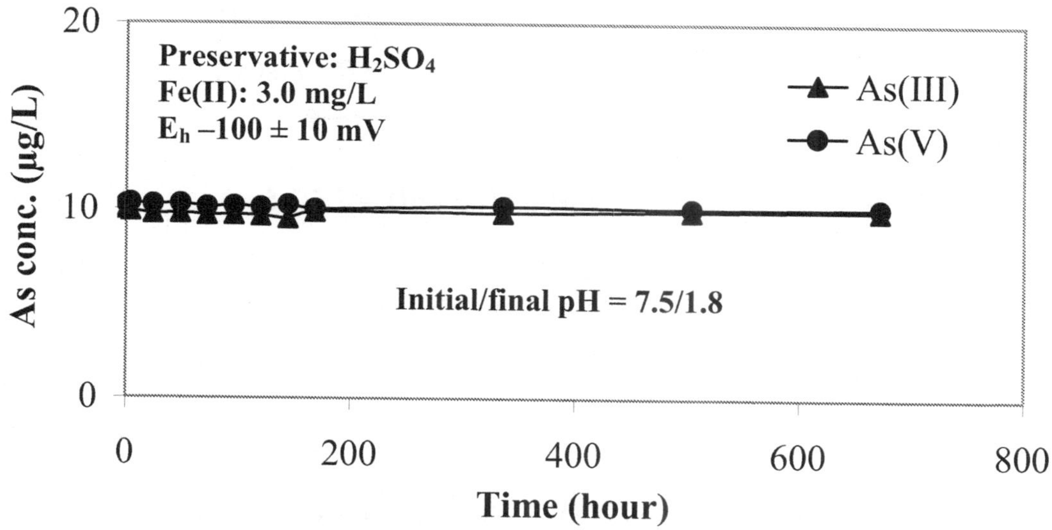

Figure 3.20 Stability of arsenic species in NSF 2 challenge water (initial/final pH 7.5/1.8) containing 3 mg/L Fe(II) with Eh −100 ± 10 mV using H_2SO_4 as a preservative in a brown PP bottle at room temperature. The test solution initially contained 10 µg/L As(III) and 10 µg/L As(V).

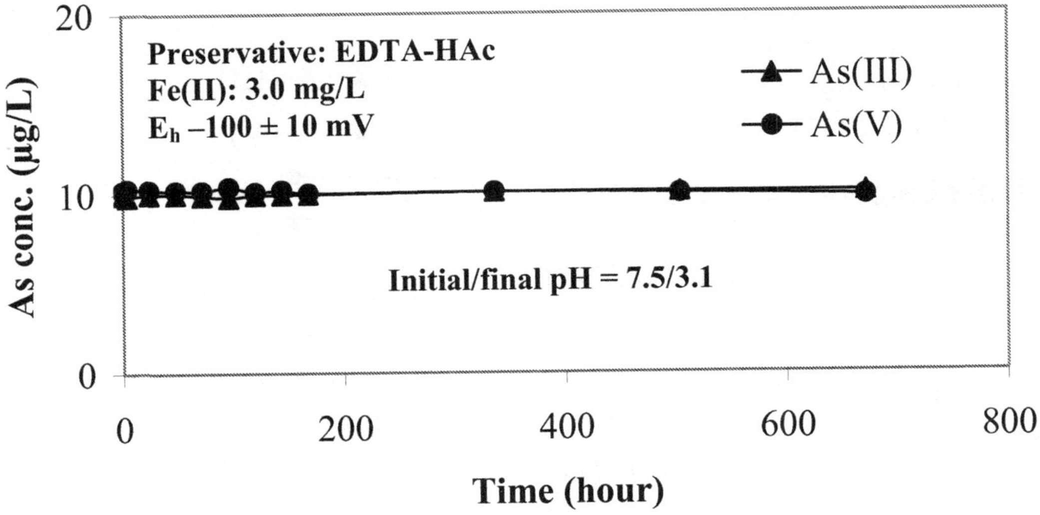

Figure 3.21 Stability of arsenic species in NSF 2 challenge water (initial/final pH 7.4/3.1) containing 3 mg/L Fe(II) with Eh −100 ± 10 mV using EDTA-HAc as a preservative in a brown PP bottle at room temperature. The test solution initially contained 10 µg/L As(III) and 10 µg/L As(V).

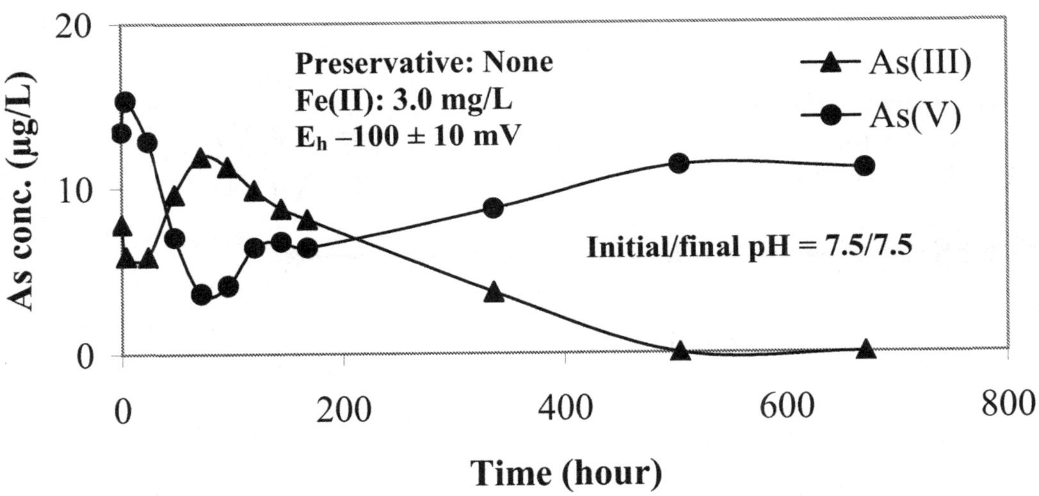

Figure 3.22 Stability of arsenic species in NSF 2 challenge water (initial/final pH 7.5/7.5) containing 3 mg/L Fe(II) with Eh −100 ± 10 mV using no preservative in a brown PP bottle at room temperature. The test solution initially contained 10 µg/L As(III) and 10 µg/L As(V).

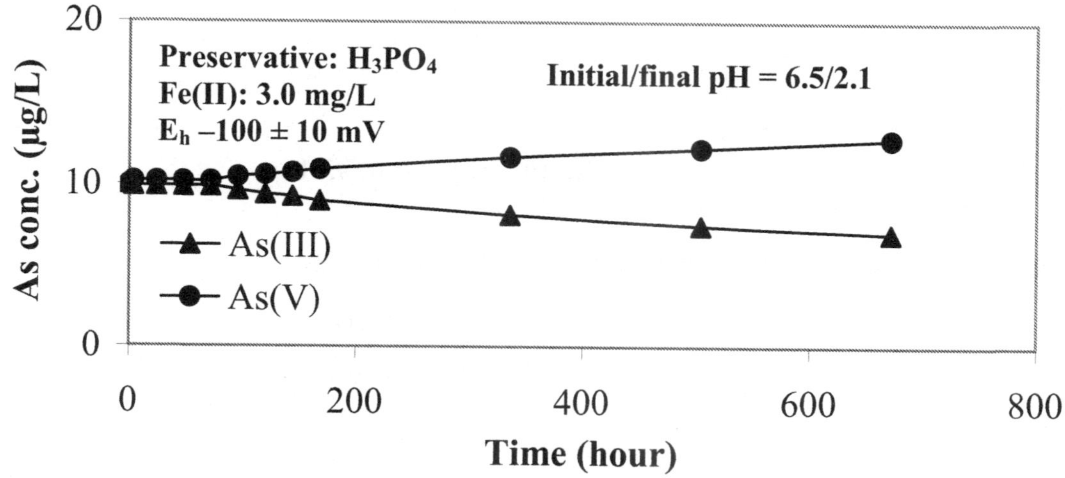

Figure 3.23 Stability of arsenic species in NSF 2 challenge water (initial/final pH 6.5/2.1) containing 3 mg/L Fe(II) with Eh –100 ± 10 mV using H_3PO_4 as a preservative in a brown PP bottle at room temperature. The test solution initially contained 10 µg/L As(III) and 10 µg/L As(V).

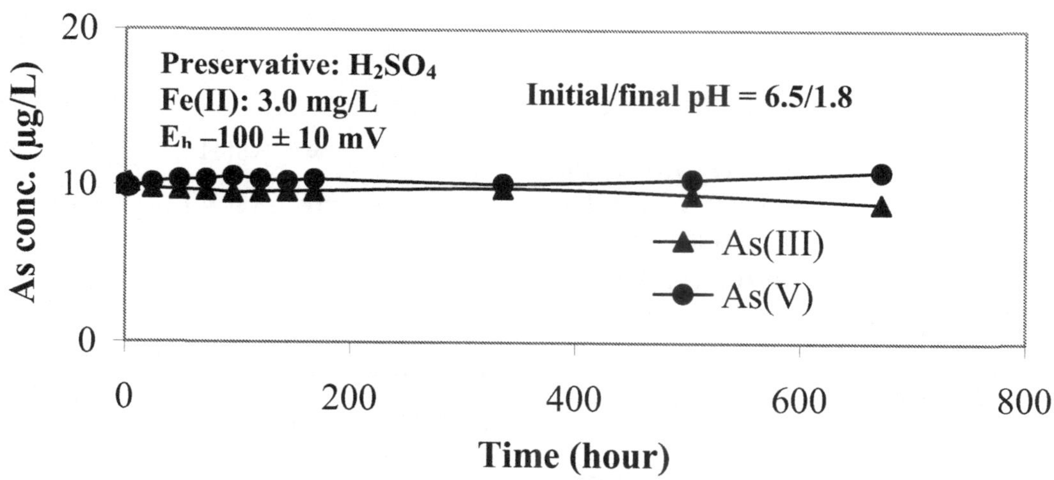

Figure 3.24 Stability of arsenic species in NSF 2 challenge water (initial/final pH 6.5/1.8) containing 3 mg/L Fe(II) with Eh –100 ± 10 mV using H_2SO_4 as a preservative in a brown PP bottle at room temperature. The test solution initially contained 10 µg/L As(III) and 10 µg/L As(V).

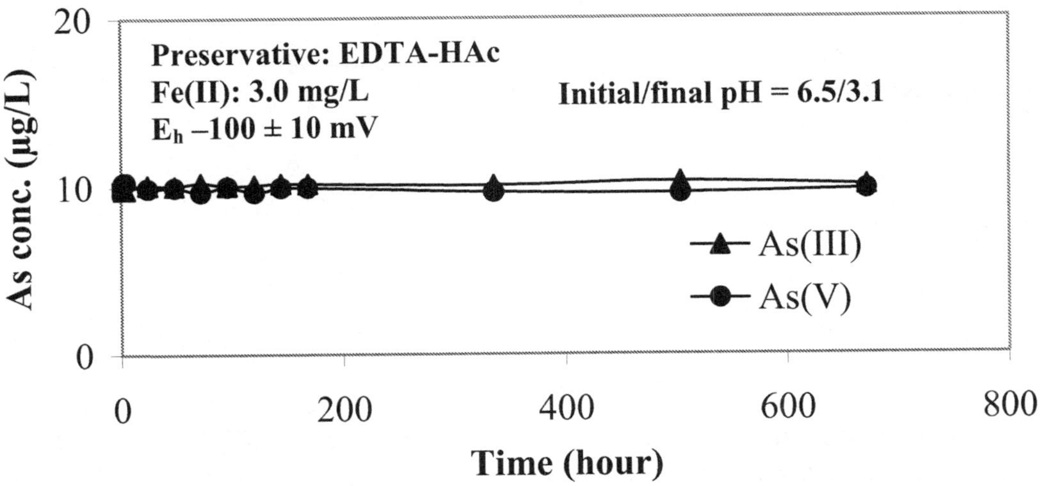

Figure 3.25 Stability of arsenic species in NSF 2 challenge water (initial/final pH 6.5/3.1) containing 3 mg/L Fe(II) with Eh –100 ± 10 mV using EDTA-HAc as a preservative in a brown PP bottle at room temperature. The test solution initially contained As(III) and 10 µg/L As(V).

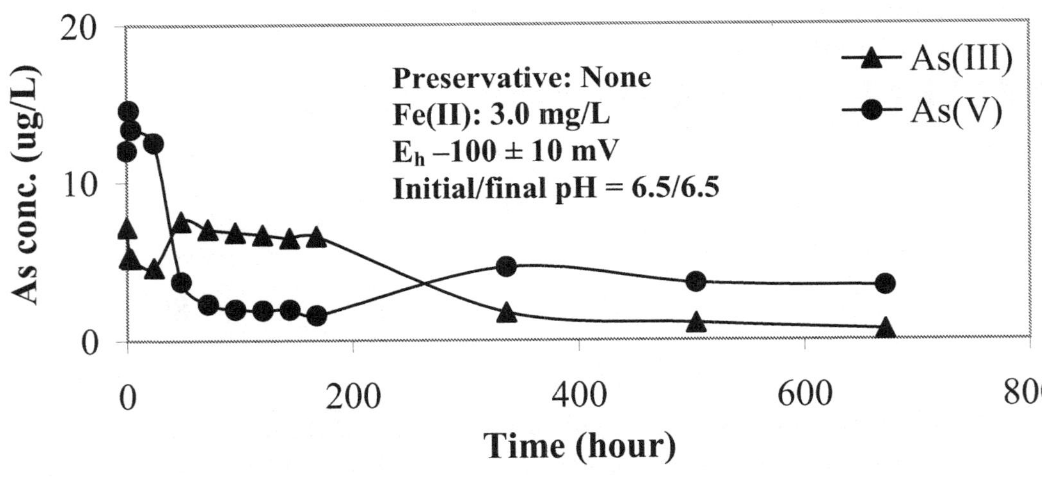

Figure 3.26 Stability of arsenic species in NSF 2 challenge water (initial/final pH 6.5/6.5) containing 3 mg/L Fe(II) with Eh –100 ± 10 mV using no preservative in a brown PP bottle at room temperature. The test solution initially contained 10 µg/L As(III) and 10 µg/L As(V).

Preservation of Arsenic Species under Oxidizing Conditions

Similar experiments to determine the stability of arsenic species with these three preservatives were carried out at the higher Eh value (+200 ± 10 mV) of the challenge water at two different pHs (8.4 and 6.5). The redox potentials of the samples (with preservatives and without preservatives) after collecting in the brown bottles were measured. The Eh values of the control sample (without preservative), and samples with H_2SO_4, H_3PO_4, and EDTA-HAc were in the range of 215-223 mV, 558-563 mV, 455 –467 mV, and 413-427 mV, respectively. As(III) and total arsenic were measured with time. The results are given in Tables 3.7-3.10 and graphically presented in Figures 3.27-3.34.

Both for H_3PO_4 (Table 3.7 and Figures 3.27 and 3.31) and H_2SO_4 (Table 3.8 and Figures 3.28 and 3.32), As(III) concentration decreased rapidly within the first few hours, and after 24 h, the As(III) concentration was almost constant. This means that H_3PO_4 and H_2SO_4 could not stop the oxidation of Fe(II) at high redox potential (in the presence of excess oxygen). When H_3PO_4 was used as a preservative, the final pH was 2.0 and the Eh was450 mV. Under this condition, both As(III) and As(V) can form thermodynamically favored (Figure 3.14) species such as $H_3AsO_3^0$, $H_2AsO_4^-$. In the presence of excess dissolved oxygen, As(III) gradually oxidized to the most favored As(V) species.

In the presence of H_2SO_4 the final pH was ~ 1.8 and the Eh was~ 550 mV. Under this condition, the most favored arsenic species is H_3AsO_4. Thus, from a thermodynamic point of view (see Figure 3.14), it is expected that H_2SO_4 cannot stabilize As(III) under the experimental conditions. It is also expected that the rate of oxidation of As(III) should be faster in the presence of H_2SO_4 than H_3PO_4. Our experimental data presented in Figure 3.35 shows the same observations as thermodynamically expected. Figure 3.35 clearly indicates the higher oxidation rate of As(III) in the presence of H_2SO_4 than H_3PO_4. Initially, the oxidation of As(III) was very fast in both cases, and then after 24 h, no more oxidation occurred. In the presence of measurable dissolved oxygen (approx. 0.26 mg/L), the competitive rate of complex formation of PO_4^{3-} or $SO_4^{2-}$with iron is slower than the rate of oxidation of Fe(II) but, after consumption of the excess O_2 the rate of complex formation is faster the oxidation process. The complexes stop the further oxidation and stabilize the arsenic species.

In the case of EDTA-HAc, the final pH was ~ 3.10 and the Eh was ~ 420 mV. Though under this condition, $H_2AsO_4^-$ is slightly more favored than $H_3AsO_3^o$, EDTA-HAc could prevent the oxidation of As(III). Our experimental data shows that EDTA-HAc preserved the arsenic species efficiently (4.19–8.06% deviation at pH 8.4 and 2.42–7.87% deviation at pH 6.5, Table 3.9).

In case of control samples at pH 8.4 without preservative added (Table 3.10 and Figure 3.30), more than 50% of the As(III) oxidized within few hours. On the other hand, at pH 6.5 (Table 3.8 and Figure 3.34), it took little longer time to oxidize. At pH 8.4 the rate of oxidation of Fe(II) is very fast and according to the Eh-pH diagram (Figure 3.14) at pH 8.4 and Eh ~ +200 mV As(V) [$HAsO_4^{2-}$] was the most favorable species than at pH 6.5 at the same Eh value. So, As(III) was oxidized to As(V) at pH 6.5 (Figure 3.34) at the slower rate than at pH 8.4 (Figure 3.30) and took longer time.

Table 3.7
Performance of H_3PO_4 under oxidizing condition (+200 ± 10 mV) at different pHs.

Initial pH	Time, h	As(III) conc. µg/L (n= 3)	Deviation (%)
	0.05	9.52	-
	2	8.28	13.02*
	4	7.36	22.69*
	24	5.59	41.28*
	48	5.47	42.54*
	72	5.74	39.71*
8.4	96	5.38	43.49*
	120	5.43	42.96*
	144	5.27	44.64*
	168	5.32	44.12*
	336	5.23	45.06*
	504	5.02	47.27*
	672	4.81	49.47*
	0.05	9.75	-
	2	9.05	7.18
	4	8.46	13.23*
	24	7.18	26.36*
	48	7.46	23.49*
	72	7.02	28.00*
6.5	96	6.77	30.56*
	120	6.89	29.33*
	144	6.78	30.46*
	168	7.16	26.56*
	336	6.89	29.33*
	504	6.75	30.77*
	672	6.81	30.15*

*indicates that the preservative failed because the deviation exceeded ± 10%.

Table 3.8
Performance of H_2SO_4 under oxidizing condition (+200 ± 10 mV) at different pHs.

Initial pH	Time, h	As(III) conc. µg/L (n= 3)	Deviation (%)
	0.05	9.54	-
	2	7.27	23.79*
	4	5.96	37.53*
	24	4.97	47.90*
	48	4.39	53.98*
	72	4.69	50.84*
8.4	96	4.46	53.25*
	120	4.55	52.23*
	144	4.49	52.93*
	168	4.49	52.94*
	336	4.47	53.14*
	504	4.49	52.94*
	672	4.46	53.23*
	0.05	9.42	-
	2	8.73	7.32
	4	8.13	13.69*
	24	6.20	34.18*
	48	6.61	29.83*
	72	6.31	33.01*
6.5	96	5.94	36.94*
	120	6.07	35.56*
	144	5.88	37.58*
	168	6.21	34.08*
	336	6.21	34.08*
	504	6.11	35.14*
	672	6.11	35.14*

*indicates that the preservative failed because the deviation exceeded ± 10%.

Table 3.9
Performance of EDTA-HAc under oxidizing condition (+200 ± 10 mV) at different pHs.

Initial pH	Time, h	As(III) conc. μg/L (n= 3)	Deviation (%)
8.4	0.05	9.88	-
	2	9.61	2.73
	4	9.46	4.25
	24	9.41	4.76
	48	9.42	4.66
	72	9.46	4.25
	96	9.43	4.55
	120	9.42	4.66
	144	9.36	5.26
	168	9.34	5.46
	336	9.40	4.86
	504	9.38	5.06
	672	9.45	4.35
6.5	0.05	9.93	-
	2	9.76	1.72
	4	9.71	2.22
	24	9.58	3.52
	48	9.78	1.51
	72	9.69	2.42
	96	9.67	2.62
	120	9.63	3.02
	144	9.52	4.12
	168	9.64	2.92
	336	9.68	2.52
	504	9.66	2.72
	672	9.63	3.02

Table 3.10
Performance without preservative under oxidizing condition ($+200 \pm 10$ mV) at different pHs.

Initial pH	Time, h	As(III) conc. µg/L (n= 3)	Deviation (%)
	0.05	8.82	-
	2	5.30	39.91*
	4	4.75	46.14*
	24	4.30	51.25*
	48	4.01	54.54*
	72	3.82	56.68*
8.4	96	3.21	63.61*
	120	2.77	68.59*
	144	2.12	75.96*
	168	1.93	78.12*
	336	0.68	92.29*
	504	0	100*
	672	0	100*
	0.05	9.23	-
	2	7.50	18.74*
	4	6.89	25.35*
	24	2.61	71.72*
	48	0.48	94.79*
	72	0	100*
6.5	96	0	100*
	120	0.48	94.79*
	144	0.74	91.98*
	168	0.99	89.27*
	336	0.74	91.98*
	504	0.65	92.96*
	672	0.54	94.15*

*indicates that the preservative failed because the deviation exceeded ± 10%.

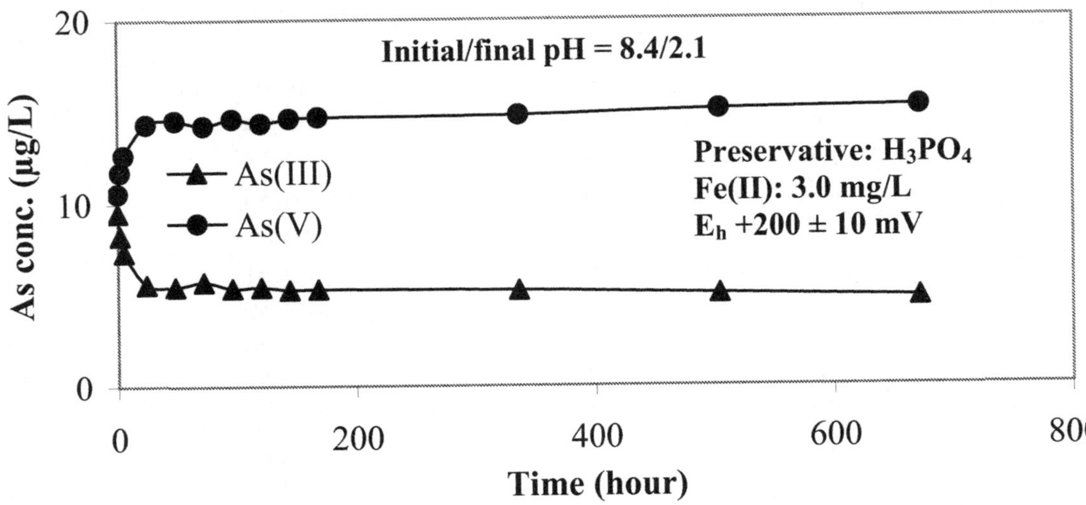

Figure 3.27 Stability of arsenic species in NSF 2 challenge water (initial/final pH 8.4/2.1) containing 3 mg/L Fe(II) with Eh +200 ± 10 mV using H_3PO_4 as a preservative in a brown PP bottle at room temperature. The test solution initially contained As(III) and 10 µg/L As(V).

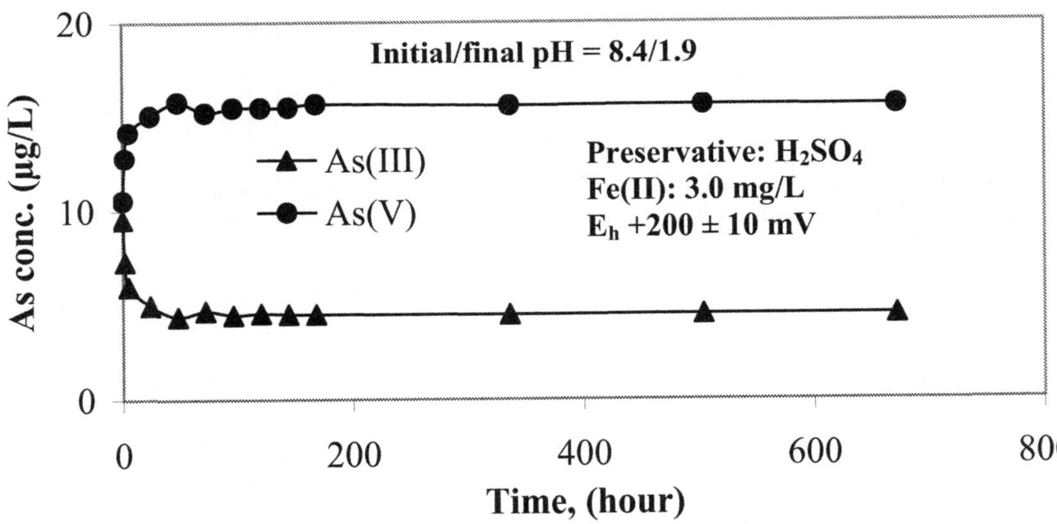

Figure 3.28 Stability of arsenic species in NSF 2 challenge water (initial/final pH 8.4/1.9) containing 3 mg/L Fe(II) with Eh +200 ± 10 mV using H_2SO_4 preservative in a brown PP bottle at room temperature. The test solution initially contained 10 µg/L As(III) and 10 µg/L As(V).

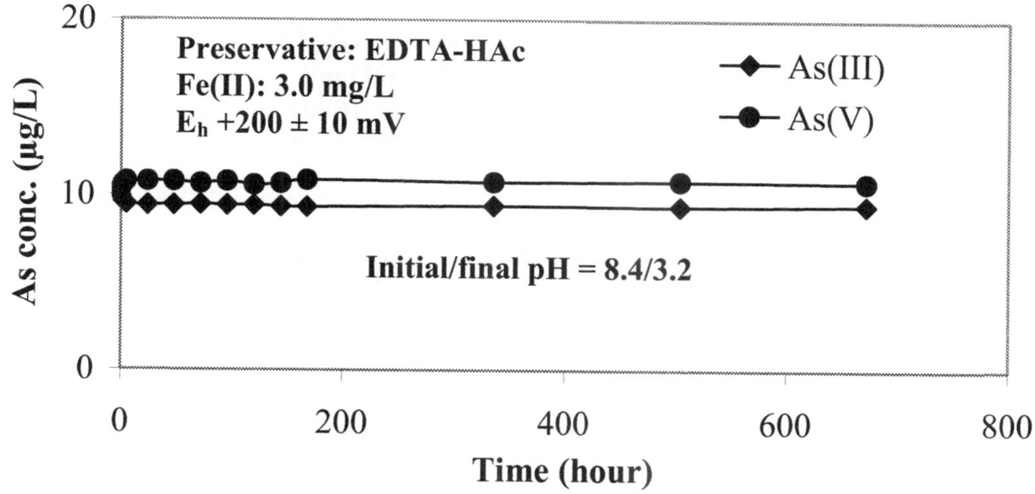

Figure 3.29 Stability of arsenic species in NSF 2 challenge water (initial/final pH 8.4/3.2) containing 3 mg/L Fe(II) with Eh +200 ± 10 mV using EDTA-HAc as a preservative in a brown PP bottle at room temperature. The test solution initially contained 10 μg/L As(III) and 10 μg/L As(V).

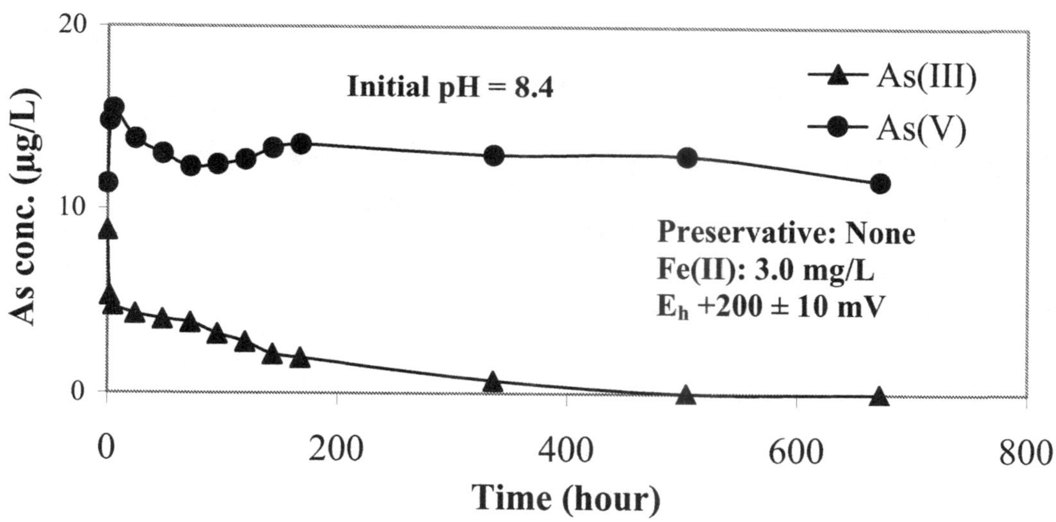

Figure 3.30 Stability of arsenic species in NSF 2 challenge water (initial/final pH 8.4/8.4) containing 3 mg/L Fe(II) with Eh +200 ± 10 mV using no preservative in brown PP bottle at room temperature. The test solution initially contained 10 μg/L As(III) and 10 μg/L As(V).

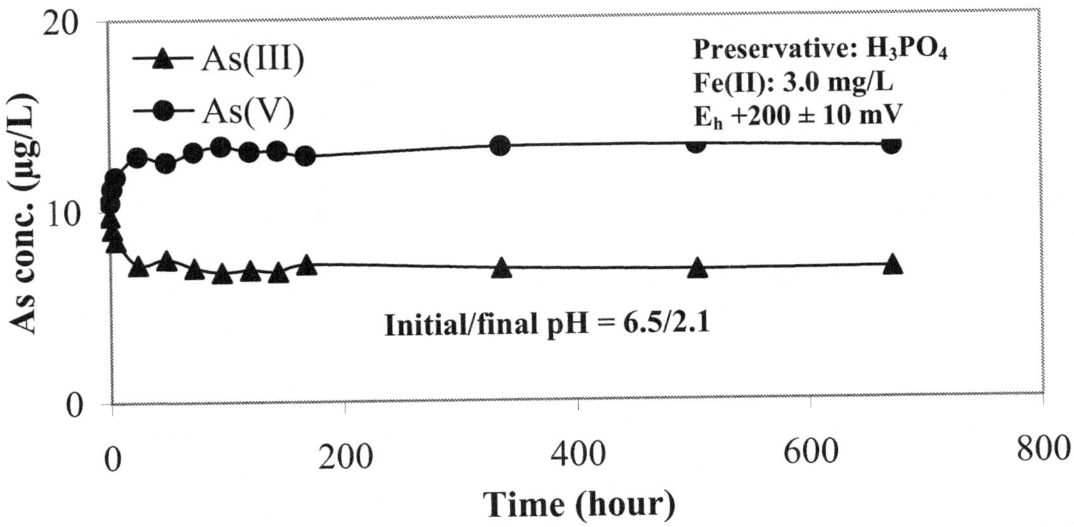

Figure 3.31 **Stability of arsenic species in NSF 2 challenge water (initial/final pH 6.5/2.1) containing 3 mg/L Fe(II) with Eh +200 ± 10 mV using H_3PO_4 as a preservative in a brown PP bottle at room temperature. The test solution initially contained 10 µg/L As(III) and 10 µg/L As(V).**

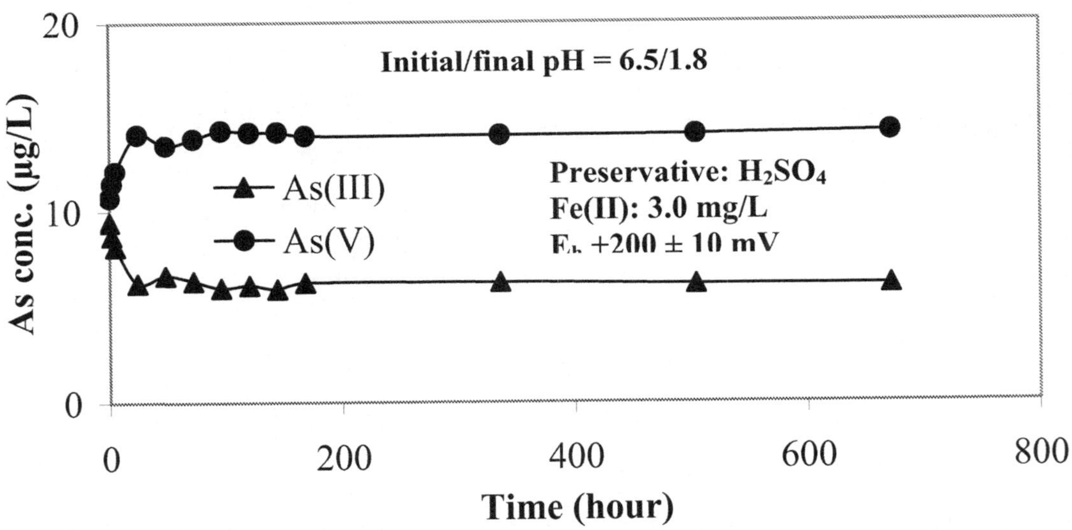

Figure 3.32 **Stability of arsenic species in NSF 2 challenge water (initial/final pH 6.5/1.8) containing 3 mg/L Fe(II) with Eh +200 ± 10 mV using H_2SO_4 preservative in a brown PP bottle at room temperature. The test solution initially contained 10 µg/L As(III) and 10 µg/L As(V).**

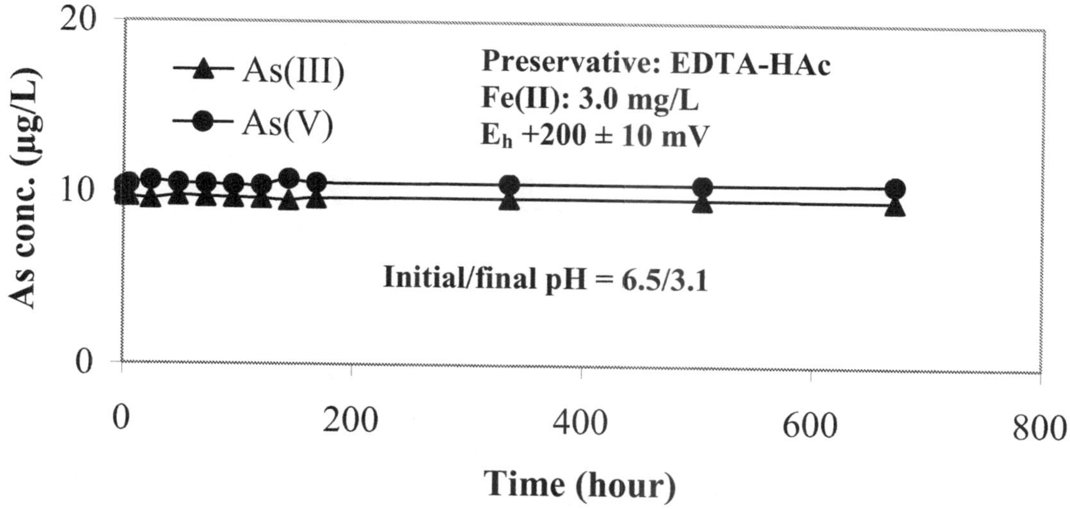

Figure 3.33 Stability of arsenic species in NSF 2 challenge water (initial/final pH 6.5/3.1) containing 3 mg/L Fe(II) with Eh +200 ± 10 mV using EDTA-HAc as a preservative in a brown PP bottle at room temperature. The test solution initially contained 10 µg/L As(III) 10 µg/L and As(V).

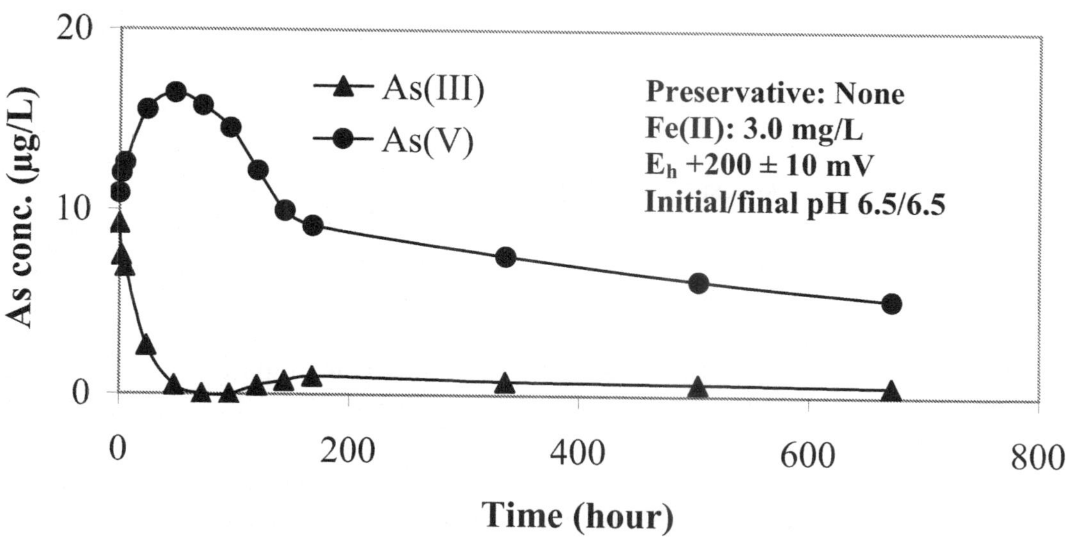

Figure 3.34 Stability of arsenic species in NSF 2 challenge water (initial/final pH 6.5/6.5) containing 3 mg/L Fe(II) with Eh +200 ± 10 mV using no preservative in a brown PP bottle at room temperature. The test solution initially contained 10 µg/L As(III) and 10 µg/L As(V).

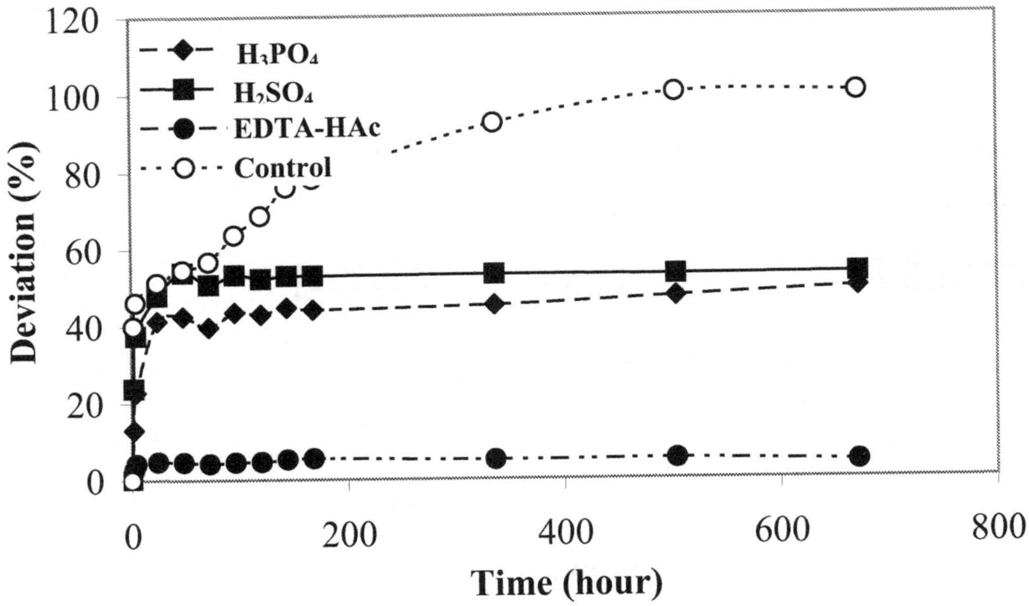

Figure 3.35 Comparison of percentage deviation of As(III) at room temperature up to 672 h (28 days) in the NSF 2 challenge water (initial pH 8.4 & Eh +200 ± 10 mV) containing 3 mg/L Fe(II) preserved with H_2SO_4, H_3PO_4, EDTA-HAc, and without any preservative in a brown PP bottle. The test solutions initially contained 10 µg/L As(III) and 10 µg/L As(V) each.

PRESERVATION AND SPECIATION OF ARSENIC SPECIES IN FIELD SAMPLES

Field Preservation Method

Based on the laboratory test results for stability of arsenic species under various pH, UV light, Fe(II), and redox conditions, the EDTA-HAc method was chosen for further study. Before field sampling, the required amounts of EDTA and acetic acid were added to 250-mL brown (opaque) PP bottles. All the bottles had been cleaned thoroughly using 1 N HNO_3 followed by a DI water rinse. Initially, 2.0 M HAc was prepared by adding 11.5 mL of 17.4 M glacial acetic acid to DI water and making up to 100 mL. The 0.1 M EDTA was prepared by adding 3.72 g disodium salt of EDTA (FW 372.24) to DI water and making up to 100 mL. To a 250-mL pre-washed brown PP bottle, 10.8 mL of 2.0 M HAc and 3.35 mL of 0.1 M EDTA were added.

At the field-sampling site all the sample lines were flushed thoroughly, and a flowing water sample was provided through a flexible tube, typically ¼ inch diameter. The opaque bottle with the preservative was filled to a level just below its neck, being careful to avoid turbulence and air entrapment. Overflow of water was not allowed because it contained acetic acid and EDTA for pH adjustment and stabilization. After capping bottle, it was mixed thoroughly by inverting it 10 times. Within 2-3 minutes of sampling and preservation, the preserved sample could be used for the immediate field-speciation test. Preserved samples and speciated samples were placed in a sample carrier box to minimize exposure to sunlight prior to transportation to the designated laboratory.

Collection, Preservation, and Speciation of Minnesota Field Samples

Samples were collected from 48 arsenic- and iron-contaminated drinking water wells in Minnesota for preservation and immediate field speciation of inorganic arsenic species. Well depth and location information is presented in Appendix B. Single samples were collected at 37 wells and triplicate samples were collected from 11 wells. Non-refrigerated samples shipped by overnight mail were received in the laboratory within 24 to 54 h of collection in the field. As(Tot) in the preserved and field-speciated samples was measured by FI-HG-AAS using the L-cysteine reduction method, and As(III) in the preserved samples was measured by FI-HG-AAS directly in the presence of citric/citrate buffer on the day the sample was received. Then all the preserved samples were stored for analysis of As(Tot) and As(III) once a week for at least 30 days.

Out of all the samples, three samples were discarded from the preservation study. One sample had been treated with chlorine. One sample contained As(III) and As(V) below the detection limit, and another sample contained As(III) below the detection limit, but the As(Tot) concentration was 6.33 μg/L, which might have resulted from Cl_2 treatment.

To check the accuracy of the data analyzed by FI-HG-AAS, a certified natural water sample (NIST 1640) was analyzed for total arsenic. The concentration of the total arsenic in the certified natural water measured by FI-HG-AAS was 26.76 ± 0.25 μg/L (n=6) that agreed very well with the certified value (26.67 ± 0.41 μg/kg, equivalent to 26.71 μg/L).

For quality control, an inter-laboratory (UH vs. UMinn) comparison of arsenic standards was made. The objective of the comparison was to determine the accuracy of total arsenic determinations at each laboratory. At UH, we prepared 10 samples and a reagent blank by spiking As(III)/As(V) in the challenge water, which was then acidified with HNO_3. The prepared sample was then divided into two parts. One portion was shipped to University of Minnesota and the other portion was kept in our laboratory for arsenic analysis. At the University of Minnesota, nine samples including two reagent blanks (one in HNO_3 and other in EDTA-HAc) were prepared. The samples were then split, and one portion sent to UH for analysis. In the UH laboratory, FI-HG-AAS was used for analysis of total arsenic in all samples. The University of Minnesota used ICP-MS for total arsenic analysis of all samples. Neither lab knew the arsenic concentrations of samples prepared by the other lab. The range, 0-100 μg/L, was known, however. After the samples were analyzed, data were interchanged between the two labs and the accuracy of each determination was established.

The analytical results are shown in Table 3.11 where it can be seen that the FI-HG-AAS values measured at UH were in good agreement with the actual arsenic content in the samples prepared in the two different labs. Total arsenic results measured at the University of Minnesota Lab were 6-36% lower than the expected values. Analysis of the NIST 1640 sample for arsenic also showed a significantly lower value when measured by ICP-MS at the UMinn lab.

Although the UMinn ICP-MS results were low, the analytical problem could not be attributed to the use of ICP-MS. For example, all the preserved and field-speciated samples were also analyzed by ICP-MS at UH. Figure 3.36 shows the correlation between As concentrations measured at UH by FI-HG-AAS and ICP-MS. The measurements are highly correlated (Slope 1.087, Intercept 1.172, and $R^2 = 0.9932$), and a Paired t-test indicates no statistical difference between two different methods of arsenic analysis. So, to minimize analysis cost, the preserved samples were analyzed for As(Tot) and As(III) by FI-HG-AAS to study the stability of arsenic species during the 30-day preservation period.

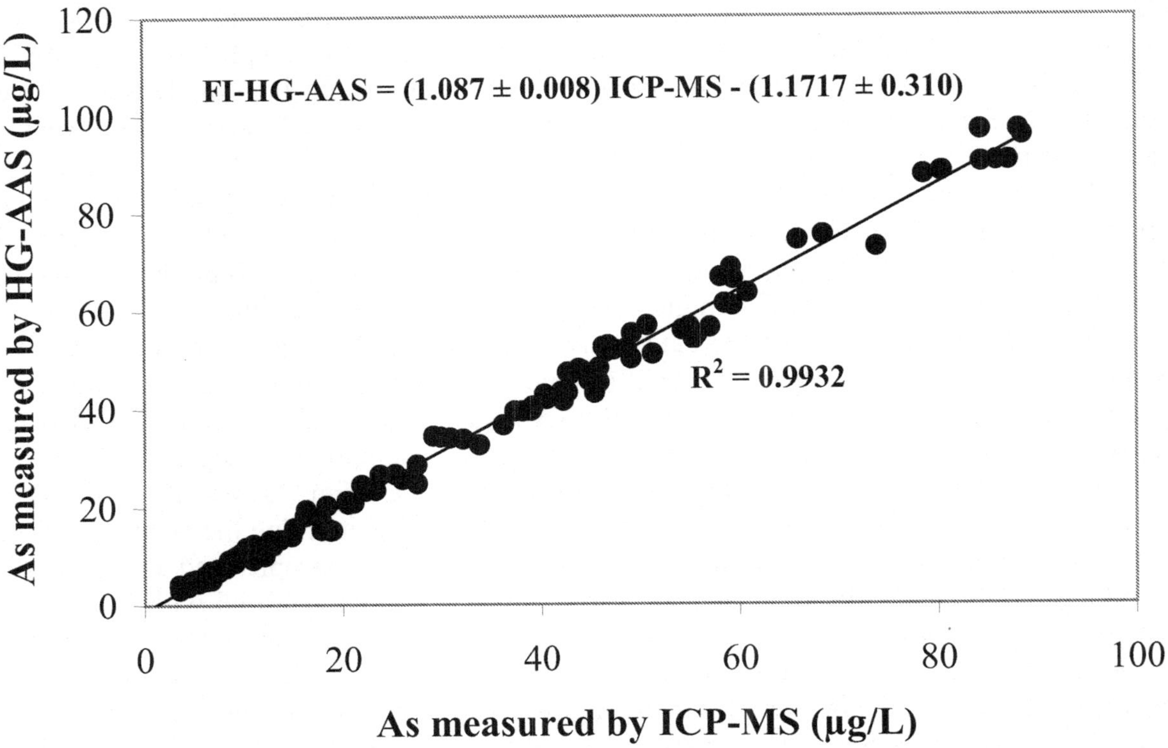

Figure 3.36 Correlation of As concentrations in preserved and field-speciated Minnesota well water samples measured by FI-HG-AAS and ICP-MS

Table 3.11
Comparison of arsenic concentrations measured by University of Houston (UH) and University of Minnesota (UMinn)

Sample ID	Prepared at	Actual Conc. (μg/L)	UH Conc. (μg/L)	UH % Error	UMinn Conc. (μg/L)	UMinn % Error
0.1 N HNO_3 As test soln A	UMinn	42.83	42.57	0.6	29.95	30.1
0.1 N HNO_3 As test soln B	UMinn	10.74	10.78	0.4	7.59	29.3
0.1 N HNO_3 As test soln C	UMinn	26.67	26.71	0.1	23.70	11.1
0.1 N HNO_3 As test soln D	UMinn	24.75	24.23	2.1	16.41	33.7
EDTA-HAc As test soln. A	UMinn	25.55	25.61	0.2	23.58	7.7
EDTA-HAc As test soln. B	UMinn	58.84	59.47	1.1	55.01	6.5
EDTA-HAc As test soln. C	UMinn	8.43	8.09	4.0	7.90	6.3
Sample 1	UH	10.00	10.21	2.1	9.33	6.7
Sample 2	UH	50.00	50.96	1.9	44.85	10.3
Sample 3	UH	80.00	80.96	1.2	72.81	9.0
Sample 4	UH	10.00	9.98	0.2	9.48	5.2
Sample 5	UH	50.00	51.50	3.0	44.64	10.7
Sample 6	UH	80.00	79.99	0.0	73.35	8.3
Sample 7	UH	80.00	81.70	2.1	51.07	36.2
Sample 8	UH	80.00	81.12	1.4	76.21	4.7
Sample 9	UH	26.67	26.49	0.7	23.90	10.4
NIST 1640	-	26.67(μg/g)	26.76(μg/L)	0.3	23.87	10.5

Concentrations of As(III) in the field-speciated samples, As(III) and As(Tot) in the preserved samples, and As(V) calculated from the difference between As(Tot) and As(III) were tabulated. The data are presented in Table 3.12, which also shows the ratios of As(III) and As(V) in each sample. These results indicate that the Minnesota well waters contain mostly As(III) with a few exceptions. The percentage of As(III) and As(V) in each sample is shown in Figures 3.37, 3.38, and 3.39. When As(Tot) concentrations were less than or equal to 10 μg/L, about 8% (average) of the As(Tot) was As(V) [Figure 3.37] and when As(Tot) concentrations were above 10 μg/L, about 15% of As was present as As(V) [Figure 3.38 and Figure 3.39].

Table 3.12

Concentrations of field-speciated As(III) and concentrations of As(III), As(Tot), and Fe in the preserved samples from Minnesota drinking water wells.

Sample ID	Time of 1st analysis in the lab after preservation in the field (h)	Conc. of As(III) in the field speciated sample (μg/L)	Conc. of As(III) in the preserved sample (1st analysis in lab), (μg/L)	Conc. of As(Tot) (μg/L)	Conc. of As(V) (Diff. of As(Tot)-As(III), (μg/L)	Ratio of As(III) /As(V)	Conc of Fe, mg/L
209387*	27	N/A	10.7	15.3	4.6	2.3:1	1.08
673611	48	20.9	20.8	23.5	2.6	8.2:1	3.28
409252	49	51.1	51.2	55.0	3.9	12.8:1	1.75
196987	48	9.27	9.48	10.0	0.73	12.2:1	0.93
167758	46	24.5	24.5	26.5	2.0	12.0:1	2.64
675566	54	4.65	4.82	4.82	0.17	26.9:1	0.96
213999*	51	72.9	72.7	90.2	17.3	4.2:1	4.12
476592	31	11.8	11.5	20.3	8.5	1.4:1	1.96
485717	30	14.1	14.9	15.6	1.5	9.2:1	2.76
444426	27	11.4	11.9	13.4	2.0	5.9:1	1.62

(Continued)

Table 3.12 (Continued)

Sample ID	Time of 1st analysis in the lab after preservation in the field (h)	Conc. of As(III) in the field speciated sample (µg/L)	Conc. of As(III) in the preserved sample (1st analysis in lab), (µg/L)	Conc. of As(Tot) (µg/L)	Conc. of As(V) (Diff. of As(Tot)-As(III), (µg/L)	Ratio of As(III)/As(V)	Conc of Fe, mg/L
566900	28	3.63	3.74	3.77	0.14	24.7:1	2.75
151140	52	26.7	27.0	33.8	7.1	3.8:1	4.41
670099	Sample was chlorinated prior to sampling and, therefore, not analyzed						
402603	48	6.70	7.47	7.61	0.61	11.4:1	0.89
174416	47	13.3	13.3	19.5	6.2	2.2:1	1.36
174947	51	24.8	24.4	32.6	7.8	3.2:1	2.27
151138	50	43.2	42.5	63.5	20.3	2.1:1	3.17
194365*	54	52.1	53.6	55.9	3.8	13.7:1	2.83
498791	52	66.7	66.5	74.3	7.6	8.8:1	1.75
673623	50	BDL	BDL	BDL	BDL	-	1.75
115406	49	5.21	5.27	5.31	0.10	53.7:1	1.86
115409	48	7.36	7.40	8.69	1.33	5.5:1	1.71
673624	54	5.53	5.29	6.77	1.24	4.5:1	3.52
515848	53	10.2	9.26	11.9	1.7	6.1:1	4.41
113676	52	11.2	10.7	12.3	1.1	10.8:1	3.37

(continued)

Table 3.12 (Continued)

Sample ID	Time of 1st analysis in the lab after preservation in the field (h)	Conc. of As(III) in the field speciated sample (µg/L)	Conc. of As(III) in the preserved sample (1st analysis in lab), (µg/L)	Conc. of As(Tot) (µg/L)	Conc. of As(V) (Diff. of As(Tot)-As(III), (µg/L)	Ratio of As(III)/As(V)	Conc of Fe, mg/L
117039*	50	87.7	88.1	95.7	8.0	11.0:1	2.67
673621	49	6.09	6.10	6.77	0.68	8.9:1	1.20
129838	48	15.8	15.3	17.6	1.8	8.5:1	2.09
673622*	46	9.20	9.196	9.85	0.65	14.2:1	1.32
117003	31	3.19	3.19	3.73	0.54	5.9:1	0.94
571710	30	51.6	52.6	55.2	3.6	14.5:1	2.78
494669*	29	43.4	41.7	68.6	25.2	1.7:1	1.44
520121	27	21.3	21.3	23.4	2.1	10.1:1	3.23
423436	26	18.2	18.5	20.7	2.5	7.2:1	2.16
524783	24	11.1	10.8	12.2	1.1	10.3:1	4.17
424478*	53	36.6	38.6	39.5	2.9	12.6:1	3.24
174627	51	61.3	59.2	66.3	5.0	12.2:1	2.58
431757	49	44.3	44.1	48.1	3.9	11.4:1	2.0
431791*	48	4.66	4.9	5.11	0.45	10.2:1	0.17
411999	47	36.7	35.5	39.6	2.9	12.4:1	1.06

(continued)

Table 3.12 (Continued)

Sample ID	Time of 1st analysis in the lab after preservation in the field (h)	Conc. of As(III) in the field speciated sample (µg/L)	Conc. of As(III) in the preserved sample (1st analysis in lab), (µg/L)	Conc. of As(Tot) (µg/L)	Conc. of As(V) (Diff. of As(Tot)-As(III), (µg/L)	Ratio of As(III) /As(V)	Conc of Fe, mg/L
129812*	31	46.9	45.7	52.4	5.5	8.4:1	4.64
463188	30	BDL	BDL	BDL	BDL	-	N/A
Schmitz	29	54.0	53.3	60.9	6.9	7.9:1	3.80
463181	28	BDL	BDL	6.33	6.33	-	0.44
545003	27	25.7	25.4	28.6	2.9	8.9:1	0.14
411965*	51	41.6	40.6	48.2	6.6	6.3:1	1.48
517044*	50	10.3	10.4	11.6	1.3	7.7:1	1.23
431789	49	4.14	4.42	4.98	0.84	5.0:1	N/A

*Indicates duplicates or triplicates of sample; N/A: Not Available; BDL: Below detection limit

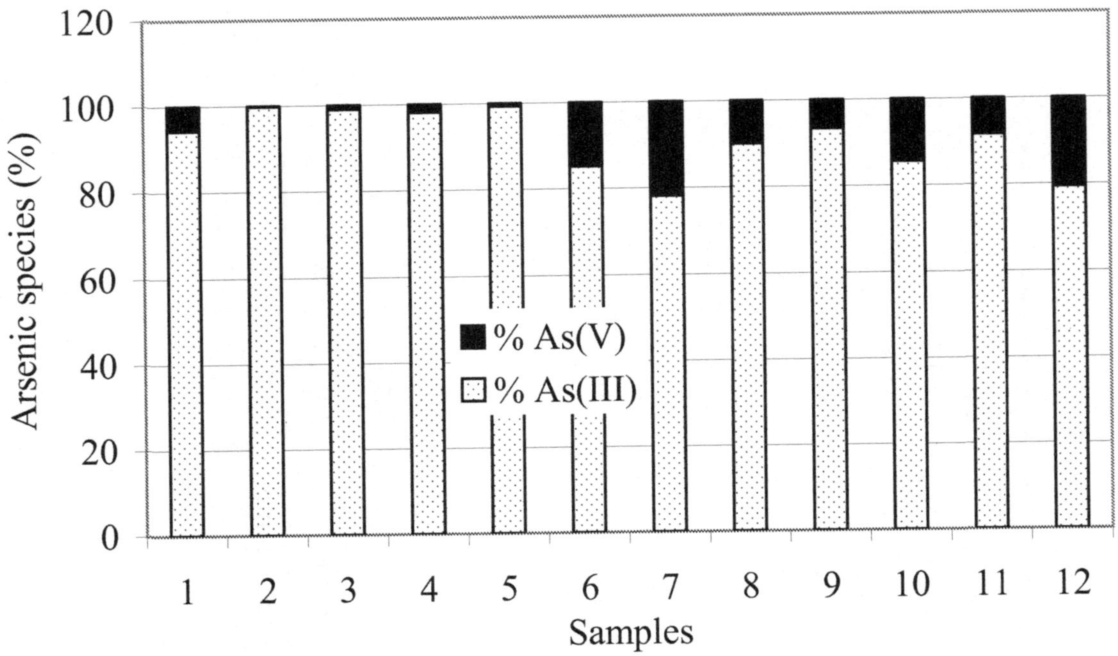

Figure 3.37 Percentage of As(III) and As(V) in the Minnesota well waters with As(Tot) concentration <10 μg/L.

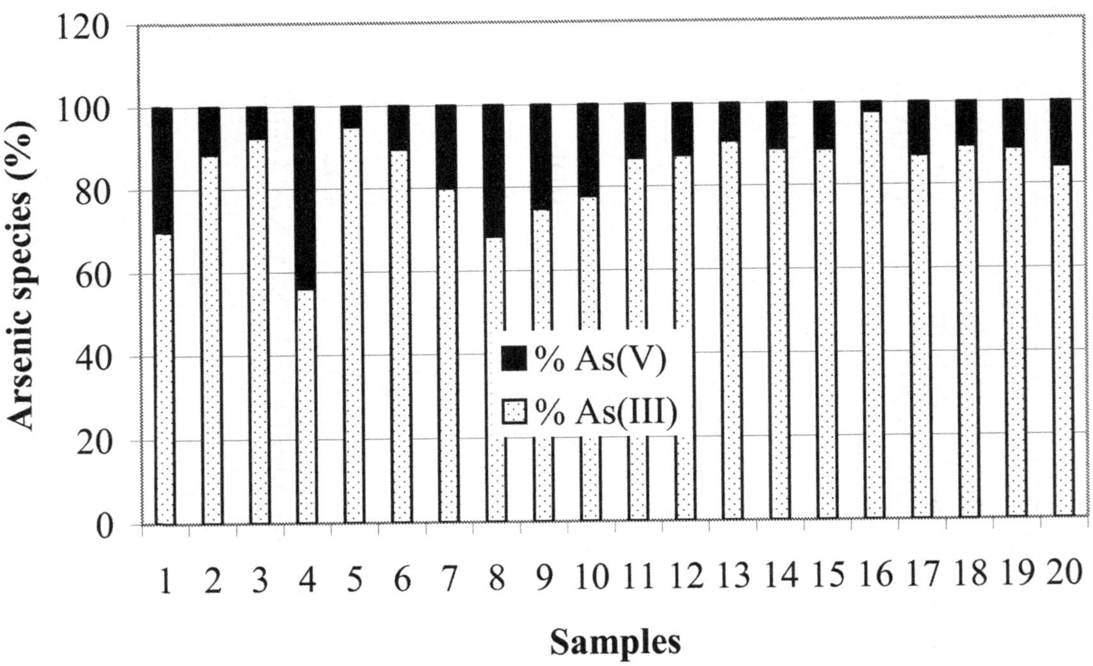

Figure 3.38 Percentage of As(III) and As(V) in the Minnesota well waters with As(Tot) concentration >10 μg/L and <50 μg/L.

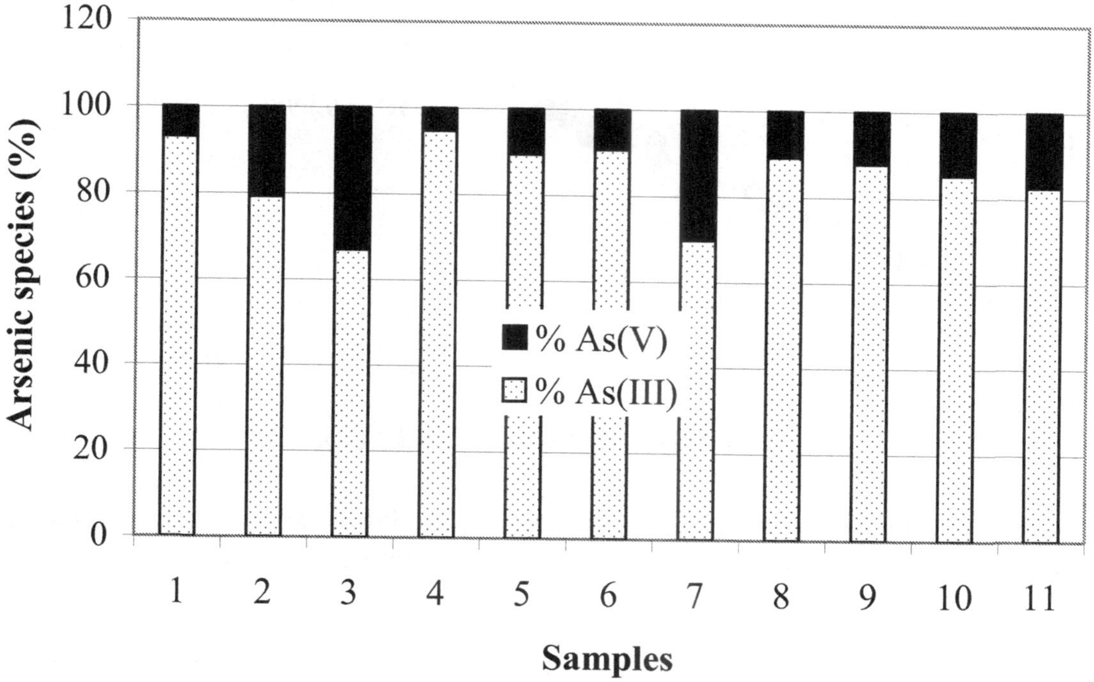

Figure 3.39 Percentage of As(III) and As(V) in the Minnesota well waters with As(Tot) concentration >50 μg/L.

Non-refrigerated samples shipped to UH by overnight mail were first speciated in the laboratory within 24 to 54 hours of collection. If the distribution of the As species had changed during transport, it could be identified by comparing the field-speciated As(III) with the As(III) concentration in the preserved samples measured in the laboratory. Figure 3.40 shows compares the As(III) result from lab speciation with the As(III) speciated in the field immediately upon sampling. These As(III) results are highly correlated and not significantly different as determined by a Paired t-test (Slope 0.9773; intercept 0.257, and $R^2 = 0.9986$). So, from Figure 3.40 based on 48 pairs of As(III) determinations on Minnesota well waters containing iron, , it can be concluded, that the EDTA-HAc preservation method maintains the As(III)/(V) distribution of preserved samples during normal transportation and storage.

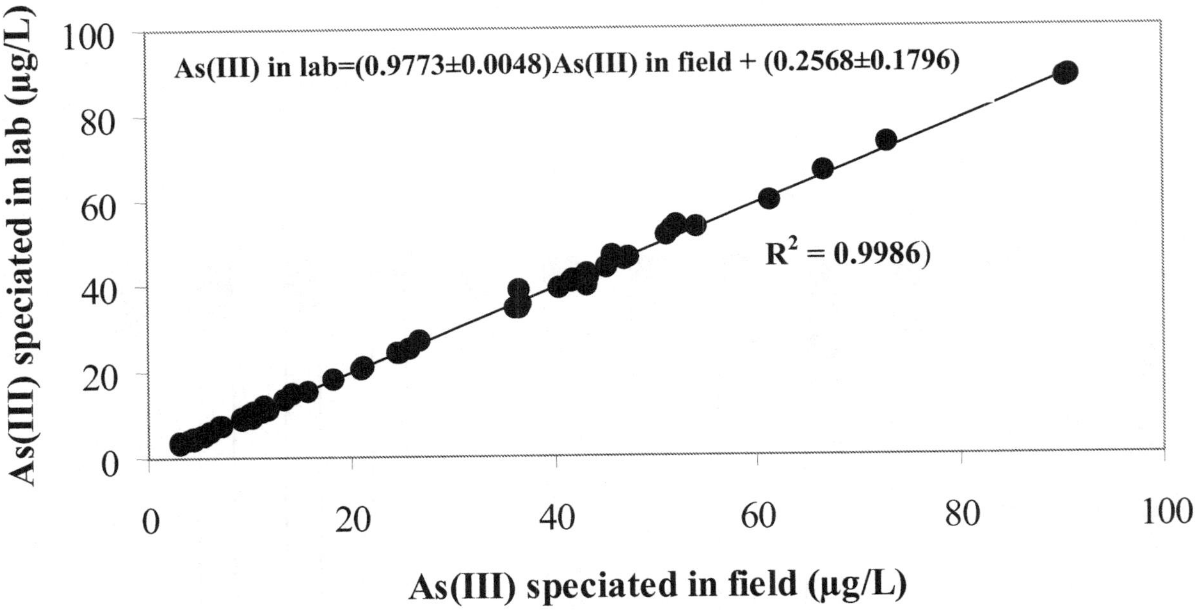

Figure 3.40. Correlation between As(III) measured in the laboratory by differential-pH hydride generation on the day it was received and As(III) determined from an immediate field speciation test.

Earlier it was demonstrated that iron plays an important role in changing the distribution of arsenic species. So, total Fe was measured in all samples by ICP-MS (See Table 3.11), and Fe concentrations varied from 0.14 to 4.64 mg/L. To see the correlation, if any, between As(Tot) and total Fe, these concentrations were plotted against one another (Figure 3.41). Although the correlation coefficient R^2 (0.0742) was not statistically significant at $p = 0.05$ level, there did appear to be a slight trend of increasing arsenic with increasing iron concentration. The increasing-arsenic-with-increasing-iron trend is more pronounced when the data plot is restricted to the eleven wells that were sampled in triplicate (Figure 3.42). This potential correlation between arsenic and iron is of some practical importance when considering arsenic treatment methods. When it holds true, less ferric iron will be necessary to precipitate the arsenic.

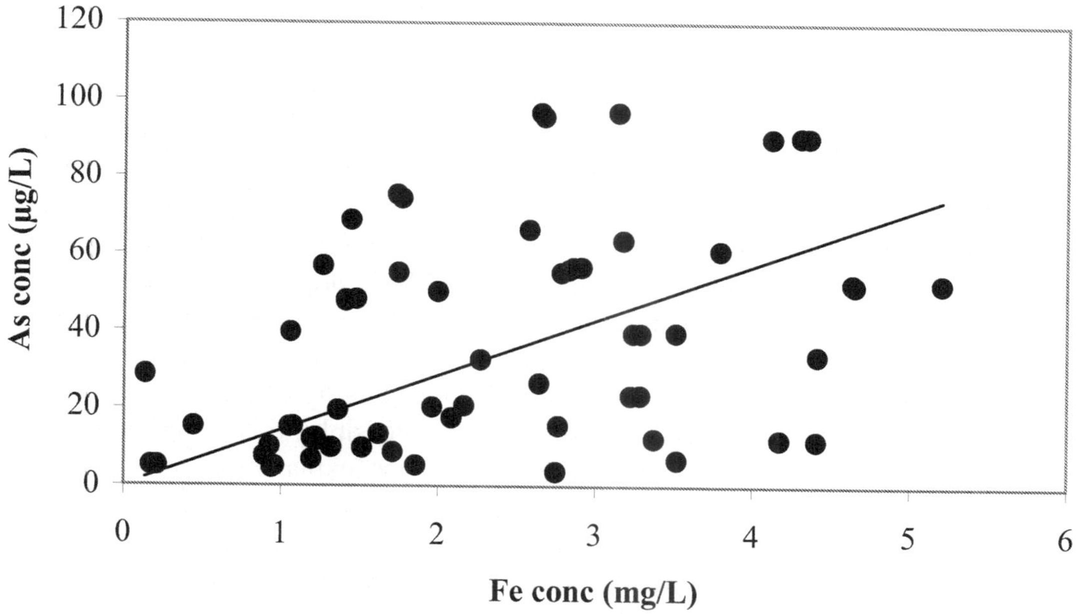

Figure 3.41. Correlation between As(Tot) and total Fe concentrations in Minnesota well waters sampled for this study.

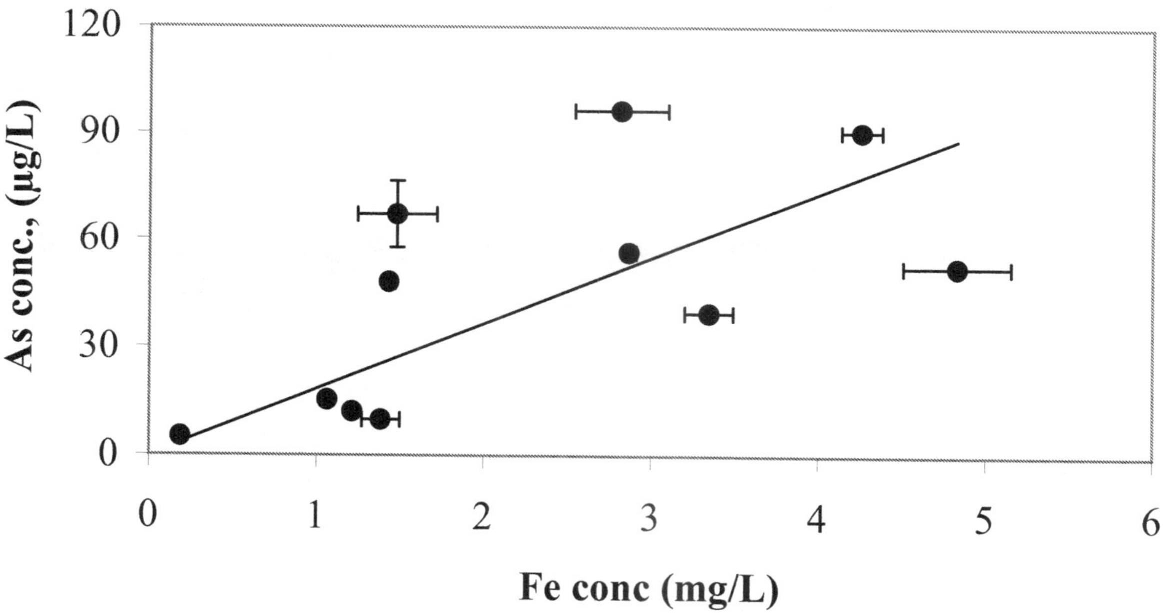

Figure 3.42 Correlation between As(Tot) and total Fe concentrations in the eleven well samples collected in triplicate for this study. The error bars along X-axis and Y-axis indicate the standard deviation in the replicate samples for Fe and As, respectively.

Thirty-Day Stability of EDTA-HAc Preserved Field Samples from Minnesota

All the 48 different well water samples (65 samples including replicates) collected and preserved in Minnesota were monitored in the laboratory for at least 30 days. The eleven replicate samples were monitored further up to 85 days. At predetermined time intervals, As(III) and As(Tot) were measured in each sample. Concentrations of As(III), As(Tot), and calculated As(V) for all samples were plotted against time where t=0 was the time of field speciation. Slopes of all trend lines for As(III) concentration measured as a function of time (up to 30 days) were calculated and plotted in Figure 3.43. These slopes should be zero if there was no change in As(III) or As(V) concentration with time. Figure 3.43 shows that slope for all samples (except one) are very close to zero, which means that the concentrations of As(III) are constant within the preservation period of 30 d. The reason why one sample significantly deviated from the zero slope (-0.011 μg/L/h) is not clear. Possibly, this sample contained a relatively strong oxidizing agent. In any case, for proper explanation and conclusion, additional samples would have to be collected and speciated over time. Redox parameters would also need to be measured. The decrease in As(III) concentration with time for this sample is shown in Figure 3.44, which demonstrates that during the first 300 h (more than 12 d) arsenic species were stable, and then As(III) slowly oxidized to As(V) and after 30 d, more than 17% As(III) was converted to As(V).

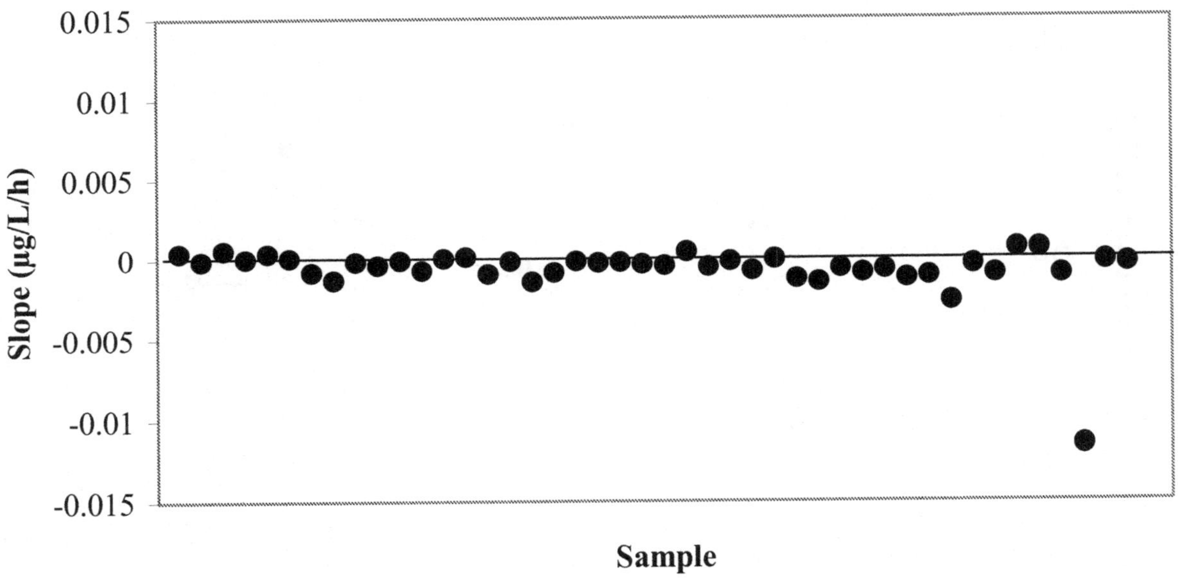

Figure 3.43 Slopes of the trend lines for As(III) measured in EDTA-HAc-preserved samples as a function of time for 30 d (considering field speciating time as zero).

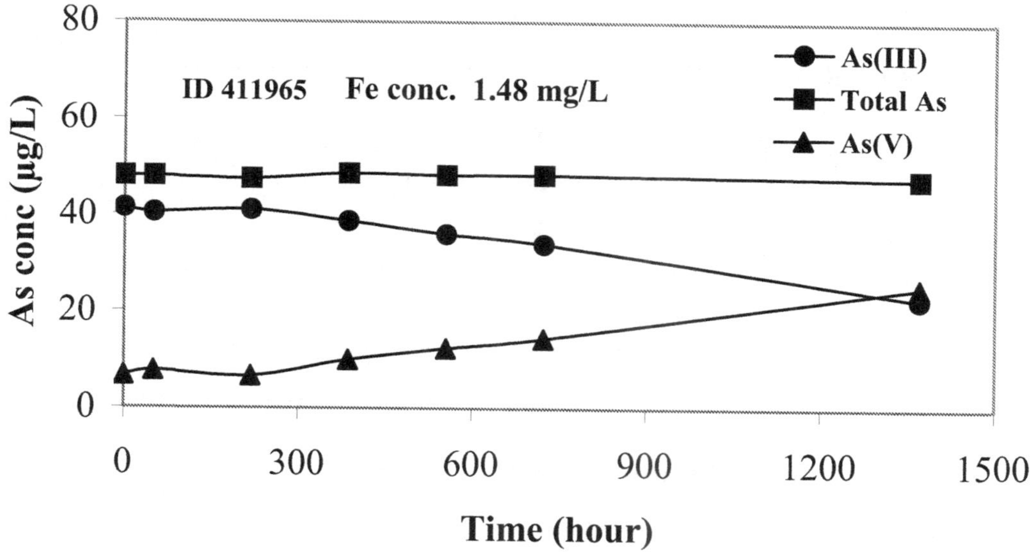

Figure 3.44 Stability of arsenic species over a period of 57 d for EDTA-HAc preserved Minnesota well water sample (ID 411965) collected in triplicate. (Error bars are too small to be visible).

The eleven samples taken and preserved in triplicate were monitored for a maximum of 85 days for arsenic speciation. Figures 3.44–3.54 show plots of wells, which are representative of the 48 wells sampled during the study. They exhibit wide variations in As and Fe concentrations. All the figures show that EDTA-HAc can preserve the arsenic species in these Minnesota well waters for more than 56 d without significant changes in their inorganic arsenic species concentrations.

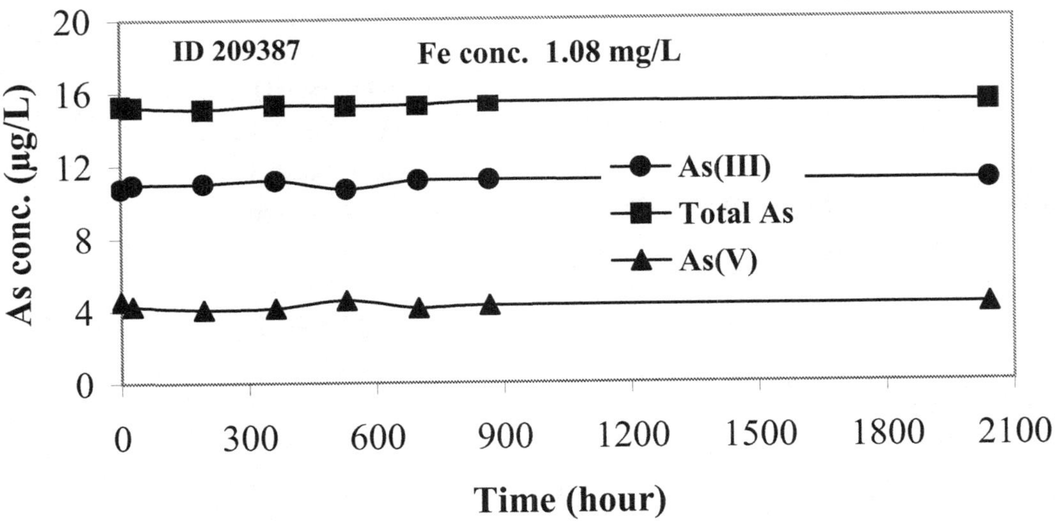

Figure 3.45 Stability of arsenic species in an EDTA-HAc-preserved sample over a period of 85 d for Minnesota well water sample (ID 209387) collected in triplicate. (Error bars are not too small to be visible).

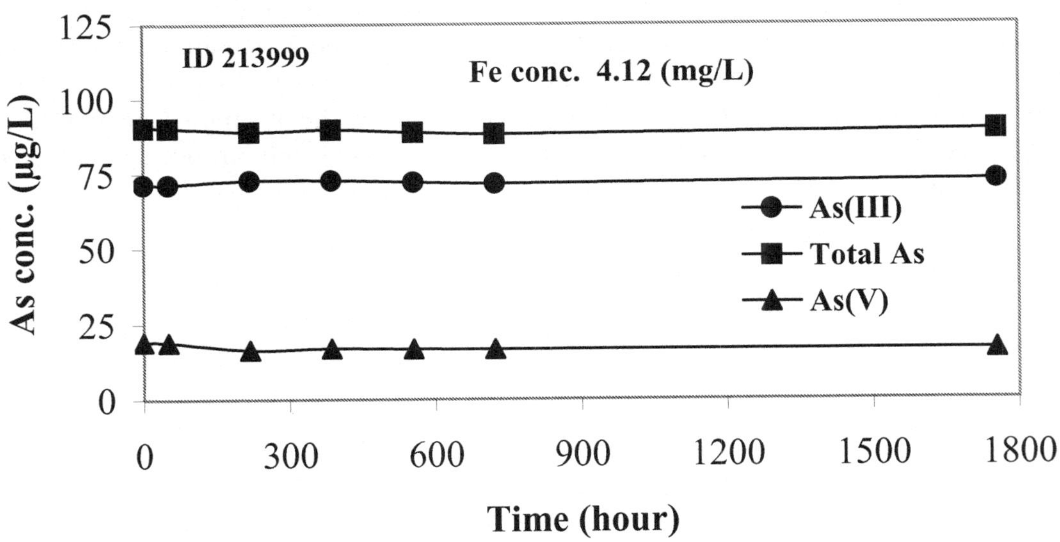

Figure 3.46 Preservation of arsenic species in an EDTA-HAc-preserved sample over a period of 73 d for Minnesota well water sample (ID 213999) collected from in triplicate. (Error bars are too small to be visible)

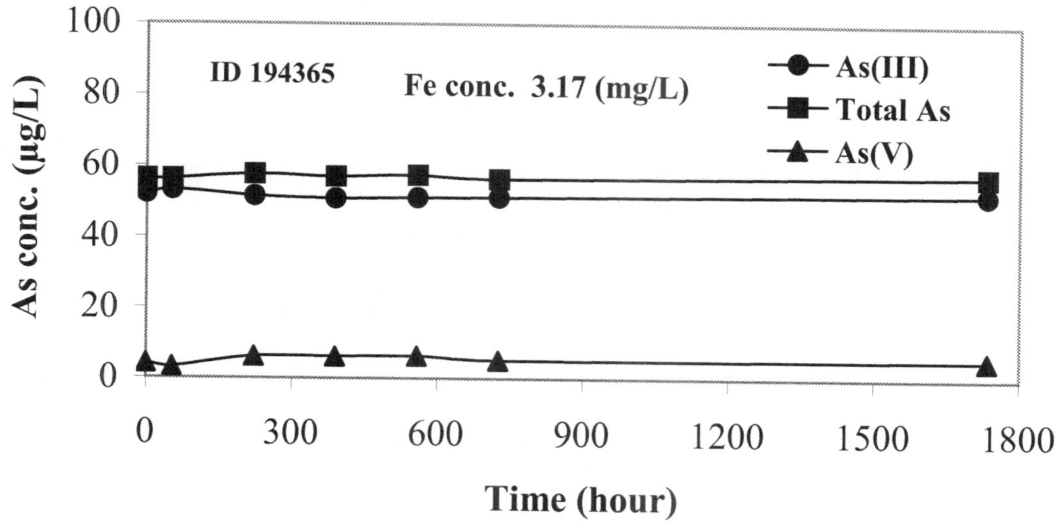

Figure 3.47 Preservation of arsenic species in an EDTA-HAc-preserved sample over a period of 72 d for Minnesota well water sample (ID 194365) collected from in triplicate. (Error bars are too small to be visible).

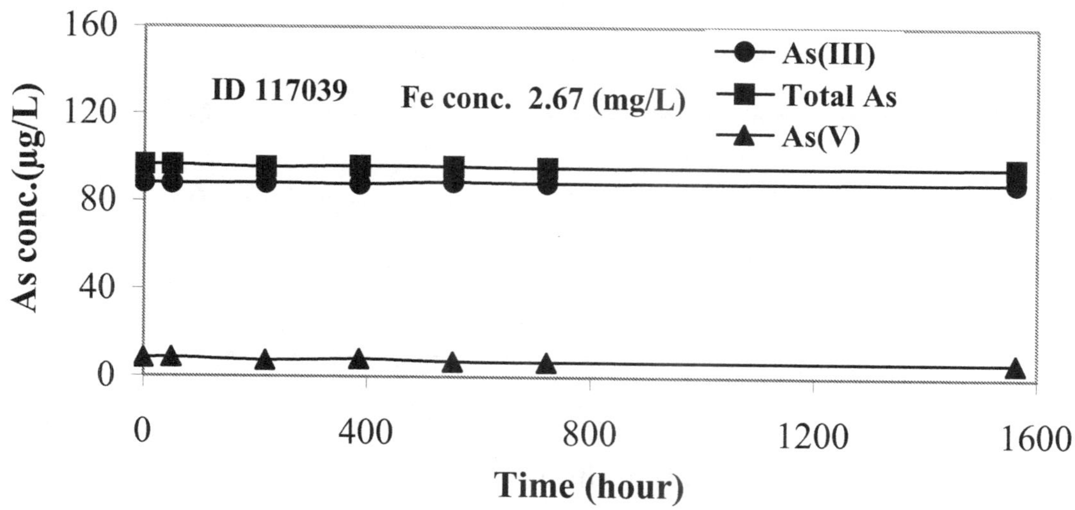

Figure 3.48 Preservation of arsenic species in an EDTA-HAc-preserved sample over a period of 69 d for Minnesota well waters sample (ID 117039) collected from in triplicate. (Error bars are too small to be visible).

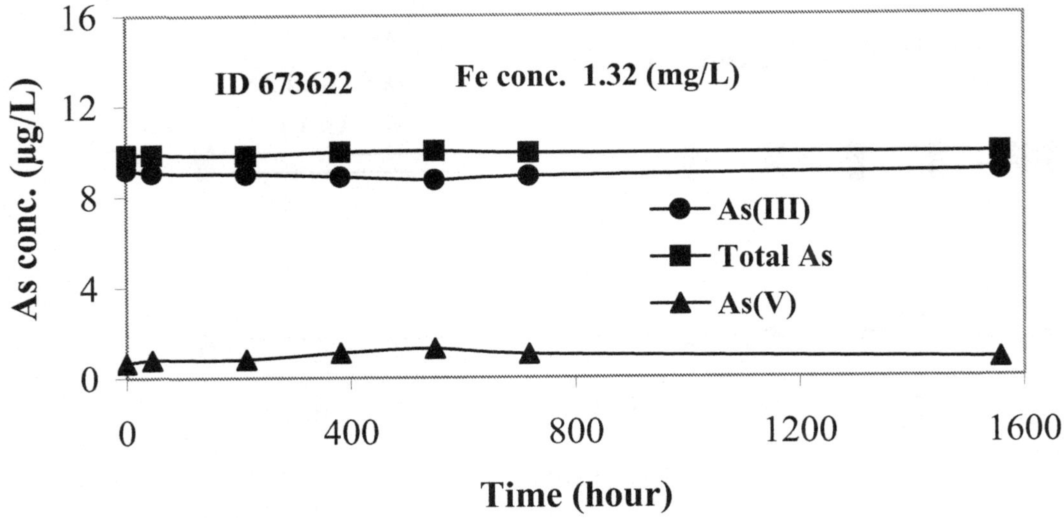

Figure 3.49 Preservation of arsenic species in an EDTA-HAc-preserved sample over a period of 65 d for Minnesota well waters sample (ID 673622) collected from in triplicate. (Error bars are too small to be visible).

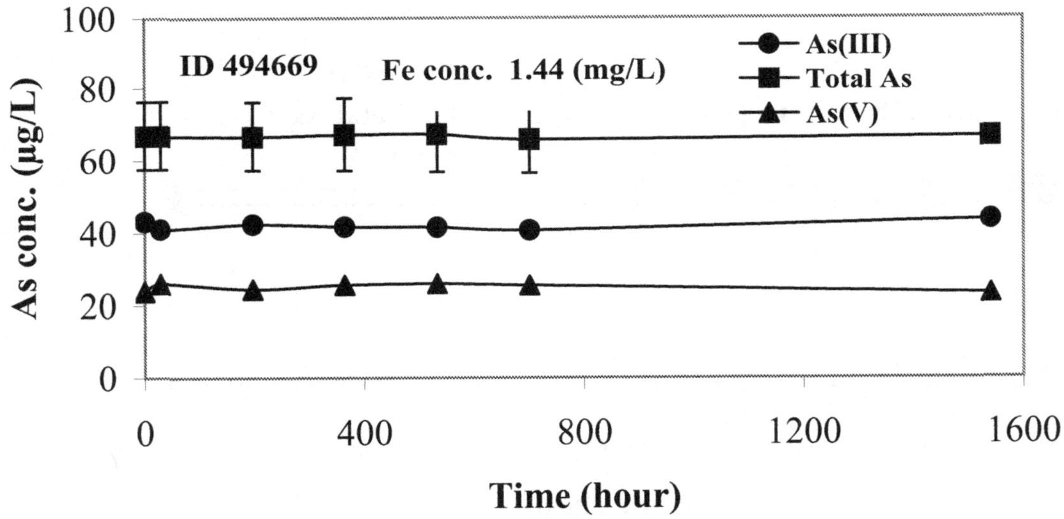

Figure 3.50 Preservation of arsenic species in an EDTA-HAc-preserved sample over a period of 64 d for Minnesota well waters sample (ID 494669) collected from in triplicate. (Error bars are too small to be visible).

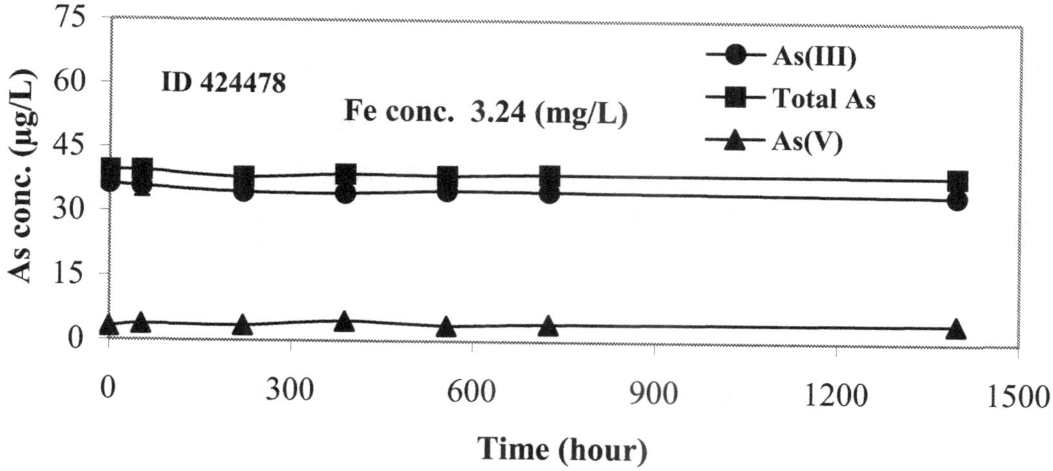

Figure 3.51 Preservation of arsenic species in an EDTA-HAc-preserved sample over a period of 30 d for Minnesota well water sample (ID 424478) collected from in triplicate. (Error bars are too small to be visible).

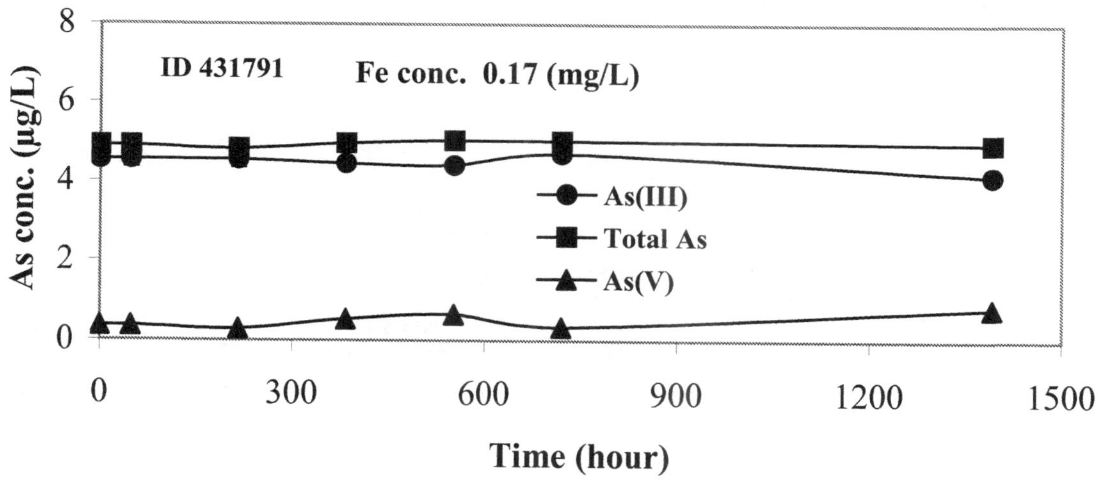

Figure 3.52 Preservation of arsenic species in an EDTA-HAc-preserved sample over a period of 58 d for Minnesota well water sample (ID 431791) collected from in triplicate. (Error bars are too small to be visible)

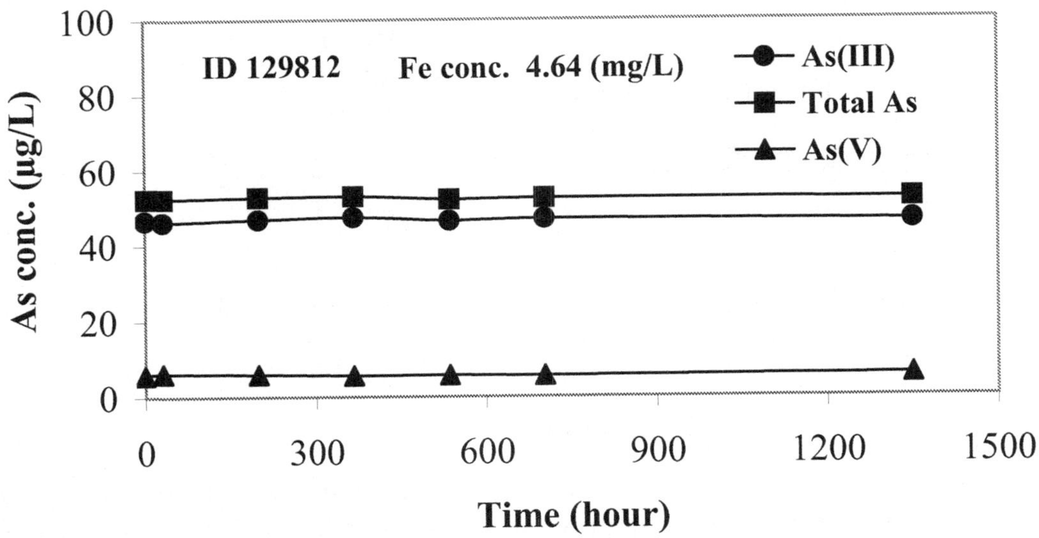

Figure 3.53 Preservation of arsenic species in an EDTA-HAc-preserved sample over a period of 56 d for Minnesota well water sample (ID 129812) collected from in triplicate. (Error bars are too small to be visible).

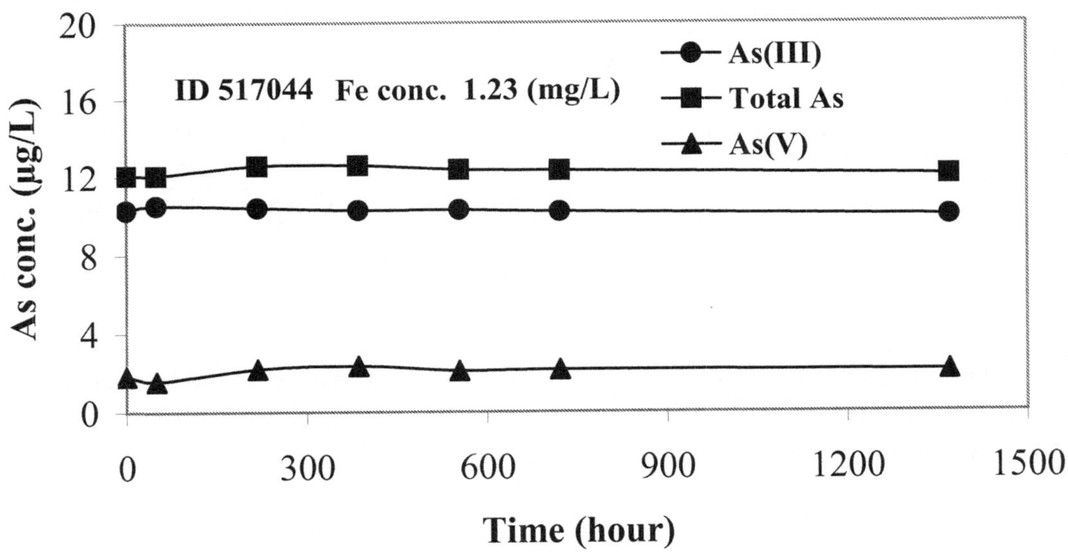

Figure 3.54 Preservation of arsenic species in an EDTA-HAc-preserved sample over a period of 57 d for Minnesota well water sample (ID 517044) collected from in triplicate. (Error bars are too small to be visible).

Field Samples from Southern California

Four samples from drinking water wells belonging to the Southern California Water Company were collected in triplicate and preserved with EDTA-HAc. All the samples were subjected to immediate field speciation to establish the inorganic As(III)/(V) ratio at the source. Following collection, preservation, and field speciation, samples were shipped to UH by overnight mail and received in the laboratory within 27 hours of collection. Field speciated and preserved samples were analyzed for As(Tot) and As(III) in the laboratory by FI-HG-AAS and ICP-MS. As was observed with the Minnesota data, the ICP-MS and FI-HG-AAS methods gave essentially the same arsenic concentration. All the Southern California Water Company samples contained very low iron. Two of the four samples contained a higher amount of As(III) and the other two contained higher amount of As(V). Table 3.13 shows the results. Because the samples were so different in their As(III)/As(V) ratios their long-term stability was of particular interest. The results of 57-day monitoring of the samples are shown in Figures 3.55 through 3.58.

Table 3.13
Concentrations of As(III) in field speciated samples compared with concentrations of As(III), and As(Tot) in the preserved samples from Southern California Water Company.

Sample ID	Age of sample when speciated in lab (h)	Conc. of As(III) in the field speciated sample (μg/L)	Conc. of As(III) in the preserved sample (in lab), (μg/L)	Conc. of As(Tot), (μg/L)	Conc. of As(V) (Diff. of As(Tot)-As(III), (μg/L)	Ratio of As(III)/As(V)
A	27	19.5±0.6 (18.75)	20.0±0.2	25.5±0.3 (24.45)	6.0±0.4	3.3:1
B	27	4.94±0.3 (4.63)	3.95±0.1	13.9±0.1 (13.41)	8.96±0.3	1:1.8
C	27	8.24±0.3 (8.07)	7.19±0.1	18.2±0.1 (18.18)	9.99±0.3	1:1.2
D	27	13.6±0.3 (13.55)	14.9±0.1	16.2±0.1 (15.97)	2.6±0.3	5.3:1

Data in parenthesis indicates the ICP-MS analysis

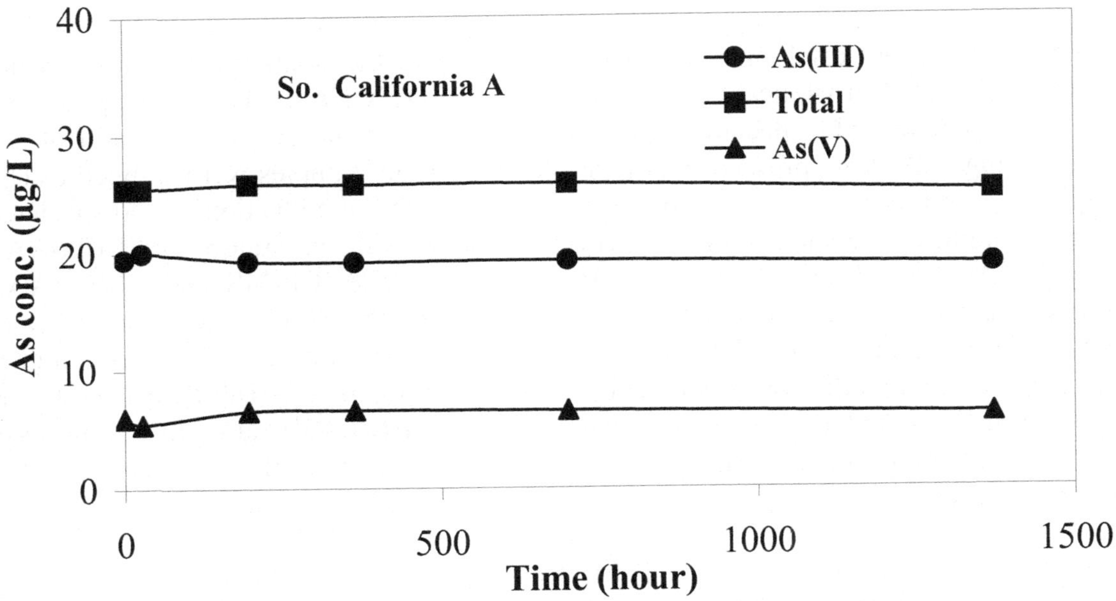

Figure 3.55 Stability of arsenic species in an EDTA-HAc-preserved sample over a period of more than 57 d for Southern California Water Company well water sample (Sample A) collected in triplicate. (Error bars are too small to be visible).

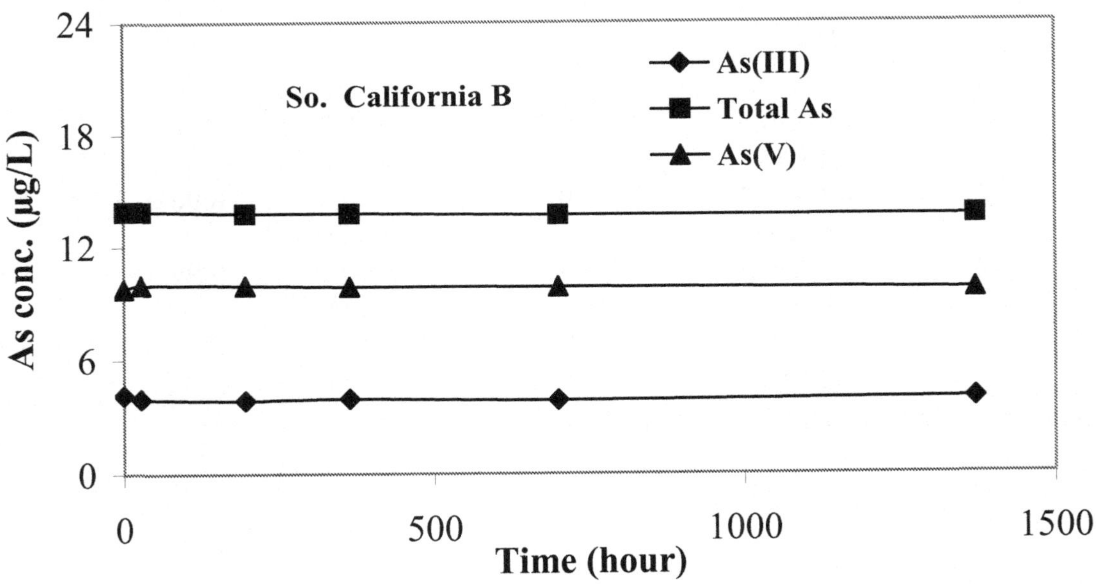

Figure 3.56 Stability of arsenic species in an EDTA-HAc-preserved sample over a period of more than 57 d for Southern California Water Company well water sample (Sample B) collected in triplicate. (Error bars are too small to be visible).

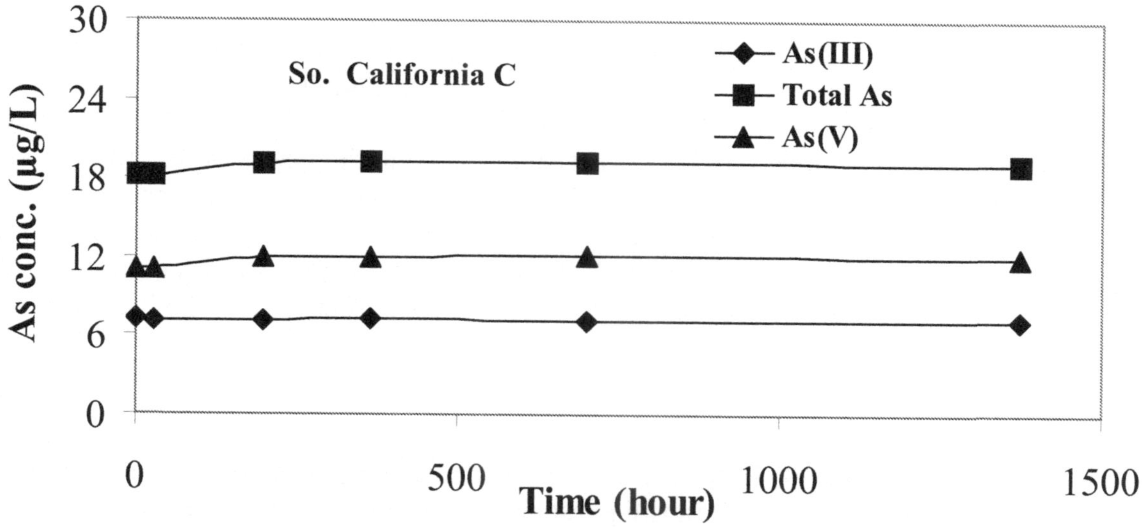

Figure 3.57 Stability of arsenic species in an EDTA-HAc-preserved sample over a period of more than 57 d for Southern California Water Company well water sample (Sample C) collected in triplicate. (Error bars are too small to be visible).

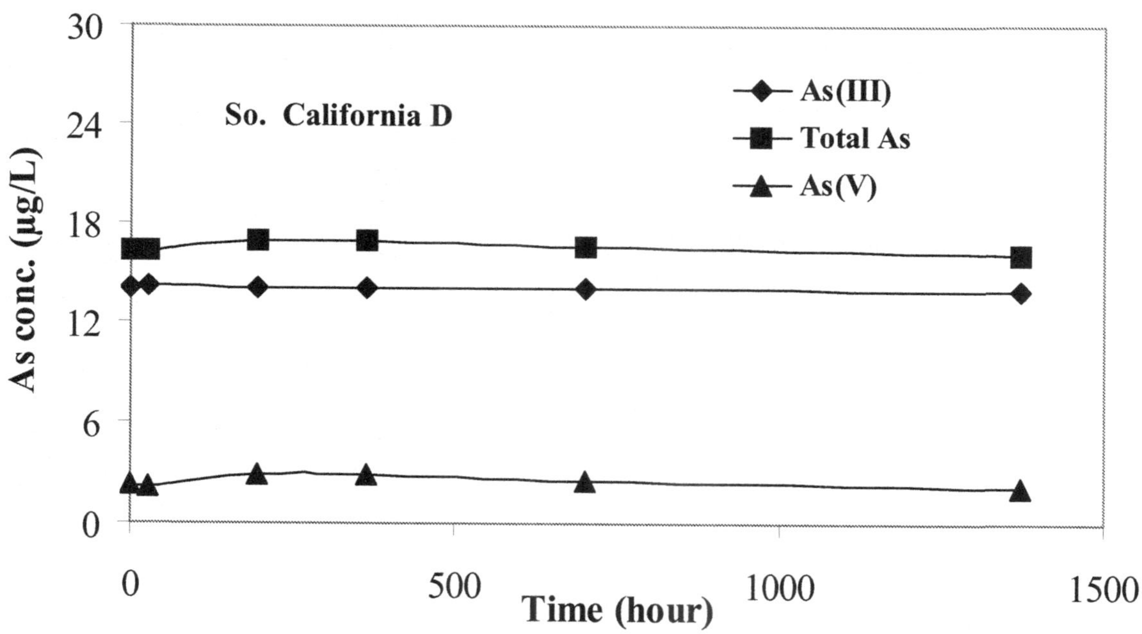

Figure 3.58 Stability of arsenic species in an EDTA-HAc-preserved sample over a period of more than 57 d for Southern California Water Company well water sample (Sample D) collected in triplicate and preserved with EDTA-HAc. (Error bars are too small to be visible).

Field Samples from North Dakota

Samples from three drinking water wells in North Dakota were collected in triplicate, preserved, and field speciated. All the preserved and speciated samples were shipped to UH by overnight mail and received in the laboratory within 54 hours of collection. Field speciated and preserved samples were analyzed for As(Tot) and As(III) in the laboratory by FI-HG-AAS after 168 h of collection. Iron was not measured in these samples. Only one of the three wells contained detectable levels of As(III). Table 3.14 shows the results of all the samples. Two samples may have been chlorinated before collection. So, only the preserved samples from the well (A) containing both As(III) and As(V) were monitored for 32 d. A result of the sample A is shown in Figure 3.59.

Table 3.14

Concentrations of As(III) in field-speciated samples compared with concentrations of As(III), and As(Tot) in the preserved samples from North Dakota.

Sample ID	Conc. of As(III) in the field speciated sample (μg/L)	Conc. of As(III) in the preserved sample (in lab), (μg/L)	Conc. of As(Tot), (μg/L)	Conc. of As(V) (Diff. of As(Tot)-As(III), (μg/L)	Ratio of As(III)/As(V)
A	12.12±0.2	11.62±0.31	13.84±0.11	1.72±0.4	7:1
B	BDL	BDL	9.04±0.16	9.04	-
C	BDL	BDL	4.96±0.28	4.96	-

BDL = Below Detection Limit

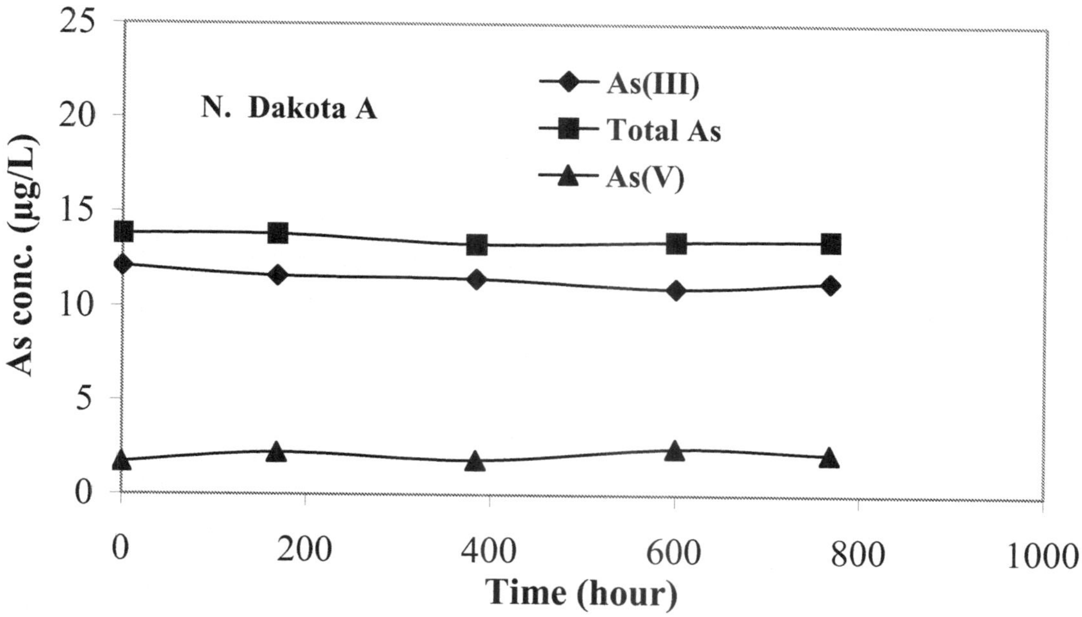

Figure 3.59 Stability of arsenic species in an EDTA-HAc-preserved sample over a period of more than 32 d for North Dakota well water sample (Sample A) collected in triplicate. (Error bars are too small to be visible).

CHAPTER 4
SUMMARY AND CONCLUSIONS, RECOMMENDATIONS, AND FUTURE RESEARCH*

SUMMARY AND CONCLUSIONS

A method for preserving the inorganic As(III)/As(V) speciation in drinking water for more than 30 days was developed at the University of Houston. The preservation-method development project was divided into seven phases: (1) preliminary experiments to evaluate the efficiency of the existing preservation methods, (2) a study of the effectiveness of known preservatives-H_2SO_4, H_3PO_4, and EDTA-HAc-in the presence absence and presence of ferrous iron, (3) a study of the effect of UV light on the arsenic species in the preserved samples using the three preservatives, (4) a study of the influence of redox potential (Eh) and pH on the preservation methods, (5) selection of the best preservation method based on the results of the laboratory experiments, and (6) field verification tests of the best preservation method — EDTA-HAc — at 55 well sites by comparing the results of immediate field speciation with lab speciation of the preserved samples over 30 days, and (7) a comparison of the FI-HG-AAS and ICP-MS methods for total arsenic analysis of the preserved samples.

This study of the preservation of inorganic arsenic species—10µg/L As(III) and 10 µg/L As(V) in a representative challenge water—resulted in the following conclusions: (Unless otherwise stated, all experiments were carried out at room temperature under laboratory light and successful preservation meant less than 10% change in As(III) concentration.)

1. In challenge water without iron, all three preservatives—H_2SO_4, H_3PO_4, and EDTA-HAc—could preserve the arsenic species for at least 7 days, the maximum duration of the test. Without preservatives, the arsenic species were stable for only one day.
2. Contrary to reports in the literature, our results indicated that EDTA alone could not preserve the arsenic species for more than twenty-four hours even in the absence of Fe(II).
3. In challenge water containing 3 mg/L Fe(II), all three preservatives — H_2SO_4, H_3PO_4, and EDTA-HAc — could preserve the arsenic species for a week or more in opaque brown polypropylene bottles under laboratory light. H_3PO_4 was an effective preservative for seven days, while H_2SO_4 and EDTA-HAc could preserve the arsenic species for at least 28 days without significant changes in their concentrations.

**** Note: This summary does not include the results of the Project Continuation Request (PCR) Report on the effects of sulfide on the EDTA-HAc preservation and speciation methods. The sulfide experiments are described in the next stand-alone chapter (Chapter 5) which has its own Introduction, Materials and Methods, Results, and Conclusions. The Executive Summary does, however, include the results of the Project Continuation Report (Chapter 5).***

4. When the same experiments with challenge water containing iron were carried out in transparent PET bottles under laboratory light, H_3PO_4 and H_2SO_4 could preserve the speciation for only 7 days while EDTA-HAc preservation was successful for 21 days. Thus, normal laboratory lighting was detrimental to preserving speciation in clear transparent plastic bottles with all three preservatives.
5. Under strong UV-A illumination (simulated direct sunlight) in the presence of Fe(II), none of the preservatives-EDTA-HAc, H_2SO_4, or H_3PO_4-could stabilize the arsenic species in clear, transparent plastic bottles for more than one hour. Under these extreme conditions, H_3PO_4 was a much better preservative than EDTA-HAc or H_2SO_4. Based on the results of the experiments with normal laboratory light and simulated direct sunlight, it is strongly recommended that samples be preserved in opaque bottles during field sample collection, transportation, and storage.
6. Under reducing conditions (Eh = -100 ± 10 mV) in the pH range of 6.5 to 8.4, and in the presence of Fe(II), EDTA-HAc could preserve the As species up to 28 days, whereas H_2SO_4 was effective for only 21 days and H_3PO_4 was effective for only 3 days at pH 6.5. Without preservatives, the speciation was immediately changed under reducing conditions.
7. Under oxidizing conditions (Eh = +200 ± 10 mV) and in the presence of >0.2 ppm DO, H_2SO_4 and H_3PO_4 could not preserve the arsenic species for more than a few minutes in opaque brown bottles. However, EDTA-HAc could preserve the arsenic species under these oxidizing conditions for at least 28 days. Some changes in speciation did occur with EDTA-HAc, but they were within the allowable 10% deviation.
8. Based on extensive experimental results in Fe(II)-contaminated challenge water, it was concluded that EDTA-HAc could be used to preserve the arsenic species for at least 28 days in opaque plastic bottles. Thus, EDTA-HAc was chosen as the preservation method for further testing. Although the alternative preservatives, H_2SO_4 and H_3PO_4, successfully preserved the original As(III)/(V) speciation under some conditions, these preservatives were generally unsuccessful for the desired 28-day period under reducing and oxidizing conditions in the sample pH range of 6.5-8.4 and in the presence of 3 mg/L Fe(II).
9. Verification tests of the EDTA-HAc preservation method were conducted at field locations in Minnesota, California, and North Dakota. Field samples were collected in bottles containing EDTA-HAc preservative and immediately speciated using the EDTA-HAc field-speciation method developed at UH under AwwaRF Project 2663. Preserved and speciated samples were shipped without refrigeration to the University of Houston by overnight mail and received within 54 hours of sampling. In cooperation with a University of Minnesota study of arsenic speciation of drinking water wells in Minnesota, samples were collected from 48 wells with 11 well samples collected in triplicate for QA/QC measurements. The Minnesota well samples contained up to 4.64 mg/L iron. With the help of cooperating utility personnel, triplicate samples were collected from four public water supply wells in California and three and three public water supply wells in North Dakota. Upon receipt at UH, all preserved samples were speciated by FI-HG-AAS using differential-pH arsine generation to distinguish As(III) from As(Tot), then As(V) was calculated by difference. The field-speciated samples were analyzed for As(III) by FI-HG-AAS and

the results were compared with the lab-speciated As(III) results on the EDTA-HAc preserved samples. These two data sets were highly correlated with a slope near 1.0 (slope 0.977; R^2 = 0.9986), which indicates that during transportation (27–55 h) the As(III)/(V) distribution in the EDTA-HAc preserved samples did not change. Further statistical analysis using the paired t-test showed no significant difference between the methods at the 95% confidence level. In the laboratory all the samples were preserved for at least 30 days and the results showed that except for one sample, all the preserved samples maintained their species distribution for at least 30 days. The triplicate samples were monitored for a total of 85 days, and no statistically significant changes were observed.

10. Because ICP-MS is so widely used for arsenic analysis—in spite of its well known chloride interference—it was compared with FI-HG-AAS in this study. The ICP-MS and FI-HG-AAS data were highly correlated (R^2 = 0.9932), however, the slope of the plot was significantly different than 1.0. (Slope = 1.08x), i.e., the ICP-MS data were consistently about 8% lower that FI-HG-AAS data. So, for routine analysis FI-HG-AAS is the more useful because it is requires less capital investment and is subject to less serious interference.
11. The main conclusion from the study was that the EDTA-HAc preservation method was capable of preserving inorganic As(III)/(V) speciation for more than 30 days in well waters under reducing or oxidizing conditions with or without ferrous iron present. Additionally, samples preserved with EDTA-HAc could be speciated immediately in the field using ion-exchange minicolumns containing strong-base anion resin in the chloride form. Results from immediate As(III)/(V) field speciation compared with lab speciation up to 30 days later were not statistically different.

RECOMMENDATIONS

The combined results of two AwwaRF arsenic analysis projects—No. 2663: Field Speciation Method for Arsenic Inorganic Species and this project: No. 2815 Preservation of Arsenic Inorganic Species—have yielded a method for immediate field-speciation and a preservation technique for speciation within 30 days of sampling. Limited field verifications of these methods have been completed, which demonstrate that they give statistically indistinguishable results. What is needed now is an inter-laboratory study of the methods to further validate their accuracy and precision. The inter-laboratory study should consist of at least the following tasks:

1. Solicit 3-5 analytical laboratories and 3-5 cooperating utilities to participate in the study. The analytical labs should be using different arsenic analysis methods such as ICP-MS and FI-HG-AAS. They should also be using different As(III)/(V) speciation techniques such as IC-ICP-MS and differential-pH hydride generation methods.
2. The central lab, which is in charge of the study, would prepare 5-7 EDTA-HAc preserved As(III)/(V) standards in the 1-100 µg/L range in challenge water, split them, and send them as unknowns to the participating analytical labs, which would speciate the samples using their preferred speciation and analysis methods.
3. The central lab would collect the data and analyze it for accuracy and precision.

4. At least 20 representative wells at the cooperating utilities would be sampled, preserved, and speciated in the field in triplicate with preserved and speciated samples sent by overnight mail to the participating laboratories.
5. The participating labs would analyze the field speciated samples and speciate the preserved samples using their chosen methods.
6. The central lab would collect the data and analyze it for accuracy and precision.
7. The central lab would prepare a report on the study and send it to AwwaRF.

FUTURE RESEARCH

The predetermined scope of work in this research project included studying the influences of ferrous iron, redox potential (Eh), pH, and strong UV light on the stability of the inorganic arsenic species drinking water samples. Based on our experimental results, we concluded that the use of EDTA-HAc in opaque bottles was the most efficient preservative to maintain the distribution of inorganic arsenic species. In the project proposal and scope of work, the influence of sulfide (S^{2-}) was not considered. However, some recent research on a project studying the intentional oxidation of arsenite by solid-phase MnO_2 media has indicated that the presence of sulfide in As(III)-containing waters may cause a loss of arsenic in some samples. Judging from our preliminary results and brief literature search, the chemistry of sulfide and As(III) is very complex. Furthermore, the presence of Fe(II), which is common in As(III)-contaminated waters, surely complicates the chemistry even more, and this has not been studied to our knowledge.

Our limited research had indicated that spiking sulfide into As(III)-contaminated water resulted in a major loss of arsenic in 0.2-µm filtered samples. For example, a 50% loss of As(III) was observed for membrane-filtered samples containing 2 mg/L sulfide subjected to the EDTA-HAc field-speciation procedure. Although sulfide was detected by odor in some of the field samples from Minnesota, a similar loss of As(III) was not observed, probably because the samples were not filtered.

All the preservation techniques studied — H_2SO_4, H_3PO_4, and EDTA-HAC — lowered the sample pH into the 1.7-3.2 range where arsenic sulfide precipitation can occur. According to the literature, with decreasing pH, the solubility of As(III) is expected to be decrease. The soluble arsenic concentration in the acid-preserved sample may be less than the true concentration due to precipitation of arsenic sulfides at low pH. Based on thermodynamic equilibrium considerations, Ferguson and Gavis (1972) reported at pH below about 5.5 and Eh values near 0.0 V, the stable solids realgar (AsS) and orpiment (As_2S_3) are formed. Experimentally, the loss of dissolved arsenic after addition of HCl and HNO_3 was also reported in the recent study of Smieja and Wilkin (2003), who carried out experiments with As(III) in the presence of S^{2-}.

As(III) may also form soluble and insoluble arsenic-sulfide complexes. Several soluble arsenic-sulfide complexes, such as di-and trithioarsenite monomers as well as dimeric and trimeric arsenic-sulfur complexes have been described or suggested during orpiment (As_2S_3) dissolution (Helz et al. 1995, Wilkin et al. 2003, Eary 1992). The presence of ferrous iron may further complicate the situation by forming the precipitates containing iron, arsenic, and sulfur, e.g., arsenopyrite, FeAsS.

In the recommended EDTA-HAc preservation method, acetic acid is used to reduce the pH to about 3.2. If sulfide is present in the water sample, the lowering of pH should affect the total dissolved arsenic and the arsenic species. According to the literature, these arsenic sulfides and arsenopyrite precipitates can redissolve forming negatively charged soluble complexes with time and during Eh and pH changes. These charged complexes would be removed by anion resins. However, is likely that we did not observe arsenic precipitation problems during field speciation and preservation, because the speciation was done within 5-10 minutes of sampling, and because the insoluble precipitates passed through the speciation columns and were subsequently measured during the hydride generation and ICP-MS procedures used for arsenic measurement. Nevertheless, because arsenic-sulfide precipitation is expected in waters containing As(III) and sulfide, it was prudent to conduct a detailed study of the possible influences of these precipitates and complexes on the speciation and preservation procedures developed under these AwwaRF studies. After this report was submitted, the UH investigators wrote a Project Continuation Request (PCR) proposal to AwwaRF justifying the need for additional funding to study the influence of sulfide on the EDTA-HAc preservation and speciation procedures. The Project Continuation Request was funded and the results of the sulfide research are reported in the following stand-alone chapter (Chapter 5) of this report.

CHAPTER 5
DETERMINATION OF INFLUENCE OF SULFIDE ON PRESERVATION AND SPECIATION OF INORGANIC ARSENIC IN DRINKING WATER*

ABSTRACT

Under AwwaRF Project 2815, Preservation of Arsenic Species, a method based on adding ethylenediaminetetraadetate (EDTA) and acetic acid (HAc) was developed for preserving inorganic As(III)/(V) species for 30 days even in the presence of ferrous iron. The EDTA-HAc preservation chemistry was previously used in AwwaRF Project 2663 for field speciation of As(III)/(V). The effect of sulfide in the sample to be speciated had not been considered in either project, so it was considered in this Project Continuation Request for Project 2815. Literature reports suggested that sulfide in the water would complex with As(III) to form negatively charged complexes that would be retained on the anion resin used to separate As(III) from As(V) and appear to be As(V). The literature also indicated that at the acidic pHs used for preservation, arsenic sulfides might precipitate thereby lowering the As(Tot) analyses. The purpose of this research was to elucidate and quantify the effects of sulfides on the Project 2815 and 2663 arsenic preservation and field speciation techniques.

The library and laboratory research conducted resulted in the following conclusions: Arsenite-sulfide chemistry in the presence of mineral acid preservatives and EDTA-HAc is extremely complex, and the rates of the reactions are slow enough that they must always be taken into consideration. In the absence of preservatives it was verified that negatively charged thioarsenite complexes exist in the waters and will prevent accurate ion-exchange speciation of As(III)/(V). When H_2SO_4 or EDTA-HAc is added as a preservative, amorphous arsenic sulfide (As_2S_3(am)) will precipitate and interfere with the determination of As(Tot). The interference with As(Tot) is worse when preserving the sample at pH 2 with H_2SO_4. It was discovered that the addition of a dilute oxidizing agent, NaOCl, causes the insoluble arsenic sulfides to redissolve and allow As(Tot) to be correctly determined.

When a ground water contains arsenic and sulfide, it is a reducing water in which all the arsenic is As(III) and no Fe(II) is present due to its insolubility in the presence of sulfide. As(V) is not stable in such a water, so As(III)/(V) speciation in the field or the lab is not an issue. However, when H_2SO_4 or EDTA-HAc field preservation-speciation is attempted on such a water, As(Tot) will likely be low and some As(V) will appear as a false positive result because of the adsorption of thioarsenite species on the ion-exchange resin. To avoid these errors it is recommended that waters containing measurable sulfide concentration not be filtered or acidified with H_2SO_4 or EDTA-HAc. Prior to As(Tot) determination, NaOCl should be added to redissolve any arsenic sulfides that may have precipitated.

Note: This additional chapter is a stand-alone report on the project continuation research conducted on AwwaRF Project 2815 —Preservation of Arsenic Species.

INTRODUCTION

The EDTA-HAc preservation method developed by University of Houston researchers under AwwaRF project 2815, Preservation of Arsenic Species, could preserve the arsenic (III)/(V) species for more than one month regardless of the Eh, pH, and Fe(II) content of the ground water sample. Speciation results measured in the laboratory on samples preserved for one month and longer were essentially identical to results from immediate field speciation tests on synthetic waters and actual ground waters from Minnesota, North Dakota, and California. The field speciation test, which also uses the EDTA-HAc preservative, was developed at the University of Houston (Clifford et al. 2004) under AwwaRF Project 2663—Field Speciation Method for Arsenic Inorganic Species. The combined results of two AwwaRF projects have yielded a method for immediate field-speciation and a preservation technique for speciation within 30 days or more of sampling.

The influence of sulfide was not included in the scope of work for either of these projects, and some recent findings at UH and in the literature suggested that sulfide may interfere with these and other field speciation and preservation techniques.

In general, concentrations of dissolved sulfide in aquatic systems are controlled by the rate of bacterial sulfate reduction and by the nature and abundance of iron in the matrix. Dissolved sulfide concentrations are able to increase only after the exhaustion of the available iron (specifically ferrous iron). In the absence of iron and under reducing environment at near neutral pH, arsenate reduction by sulfide is rapid via the formation of anionic thioarsenite species. Within the pH range 6-10, the concentrations of thioarsenite species are high when the concentration of the dissolved Fe(II) is low (Wilkin et al. 2003). So, it is expected that in the natural reducing environment where arsenic and sulfide are present, the concentration of iron will be low, and the predominant arsenic species will be thioarsenite complexes and/or arsenite. The thioarsenite species have been reported to influence the speciation and preservation of As(III)/(V) (Jay et al. 2004 and Smieja and Wilkin 2003)).

Objectives

The objective of this project continuation research was to evaluate the influence of sulfide on preservation and speciation of inorganic arsenic in drinking waters. To meet the main objective, the following specific tasks were carried out:

(1) The solubility of As(III) in EDTA-HAc preserved challenge water containing different amounts of sulfide and no ferrous iron was established. N_2-sparged challenge water with 20 μg As(III)/L was spiked with 0.5, 1.0, 2.0 and 4.0 mg S^{2-}/L and when sampled, was immediately preserved with EDTA-HAc as would occur during field sampling. Preserved samples were shaken and filtered with 0.2 μm membrane filters to remove any precipitated arsenic sulfides at 0, 5, 10, 30, 60, and 180 min and 1, 2, 5, 10, 20, 30, and 45 days. The filtrates were analyzed for As (Tot) and As(III).

(2) The solubility of As(III) in EDTA-HAc-preserved challenge water containing 2 mg/L sulfide and 2 mg Fe(II)/L was studied.

(3) If arsenic was precipitated from the preserved samples containing sulfide, in steps (1) and (2) above, attempts were made to oxidize the arsenic sulfide in the preserved samples using H_2O_2, NaOCl, peracetic acid, or Oxone.

(4) The fate of possible anionic thioarsenite species in the EDTA-HAc preserved samples without ferrous iron during immediate speciation tests using chloride-form resin minicolumn was determined. The EDTA-HAc preserved samples containing As(III) and different amounts of S^{2-} were subjected to speciation with chloride-form minicolumns at 2, 5, 10, 30, and 60 min. Speciated samples were analyzed for As(Tot).

(5) The negatively charged thioarsenite species were quantified using a Cl-form anion-exchange resin column. When negatively charged thioarsenite complexes interfered with As(III) speciation in the minicolumns, attempts were made to form neutral thioarsenite complexes using Cu(I) so as to maintain the ability of EDTA-HAc preserved samples to be speciated in the minicolumns.

(6) Based on an analysis of the data, in steps (1) – (5) above, the preservation and field speciation procedures were modified.

Significance of the Work

When arsenic and sulfide are present in a water sample, the analyses for total arsenic (As(Tot)) and As(III)/(V) speciation will be affected. If both arsenic and sulfide are present in underground waters, several compounds may form depending on the pH and the concentrations of arsenic, and sulfide. At near-neutral to alkaline pH, arsenite and sulfide are present mainly as water-soluble thioarsenite species and H_3AsO_3. As(V) is very unstable in anoxic sulfidic water and is rapidly reduced to thioarsenite species and arsenite. Therefore, one would not expect As(V) to occur in sulfidic water.

Under acidic conditions, arsenic sulfides can precipitate, thus the acidic preservation techniques in common use may interfere with speciation. It is the common practice to reduce the pH to <2 using HCl or HNO_3 to preserve the arsenic until an analysis can be performed (APHA, AWWA, and WEF 1992). In the case of preservation of arsenic species, EDTA and acetic acid (HAc) are used to complex iron lower the pH to about 3.2. Edwards et al. (Edwards et al. 1998) used H_2SO_4 and pH 2.0 for the preservation of arsenic species in waters. When HCl, HNO_3, acetic acid or H_2SO_4 is added to the sulfidic water, precipitation of arsenic sulfide occurs, and the particulate arsenic sulfide is not included in the As(Tot) analysis. The result is falsely low As(Tot) analyzed. When speciating arsenic in sulfidic water using anion-exchange resin, the presence of As(V) is falsely reported due to the adsorption of anionic thioarsenite species on the resin column (Jay et al. 2004). This will lead to incorrect treatment decisions and possibly false assessment of toxicity. Recent literature (Rader et al. 2004) shows that the toxicities of As(III) and thioarsenite species are different. Thioarsenite species are not bioavailable, and their toxicities are lower than As(III). The present work will help the water supply community to know the actual arsenic concentration and speciation in the supplied water when sulfide is present.

Review of Previous Studies

Under anaerobic conditions, sulfates and organic sulfur compounds present in ground water are reduced by bacterial activity to appreciable concentrations of sulfide, which is present in the form of H_2S or HS^- depending on the pH. In the presence of sulfide, the ground water arsenic chemistry is very complex. As(III) may form soluble and insoluble arsenic-sulfide complexes depending on the pH. The experimental data (Mironova and Zotov 1980, Mironova et al. 1984, Wilkin et al. 2003, Webster 1990), compilations of thermodynamic data (Pokrovski et al. 1996, Helz et al. 1995, Shock et al. 1997), and theoretical modeling (Helz et al. 1995, Tossel 1996, Tossel 1997, Cleverley et al. 2003) provide a basic understanding of the reactions in the As-S system. However, the distribution of the aqueous and solid As-S phases, their relative stability and stochiometry are poorly understood. In sulfidic waters, As(III) may form amorphous As_2S_3(am) or AsS(am), or water soluble thioarsenite complexes depending on the pH and the concentrations of As(III), and sulfide. At relatively low pH, As(III) may be precipitated as an amorphous arsenic sulfide, As_2S_3(am), which can undergo spontaneous solubilization with time producing different water-soluble thioarsenite complexes, such as di-and trithioarsenic monomers as well as dimeric and trimeric arsenic-sulfur complexes (Mironova and Zotov 1980, Mironova et al. 1984, Wilkin et al. 2003, Helz et al. 1995, Spycher and Reed 1989, Krupp 1990). Based on thermodynamic properties, Helz et al. 1995 rejected the formation of dimeric species of the form, $H_xAs_2S_4^{x-2}$. According to them, the trimer, $As_3S_4(SH)_2^-$, is the predominant thioarsenite species in systems saturated with As_2S_3(am), whereas in undersaturated solutions, monomers, e.g., $AsS(OH)(SH)^-$ and $AsS(SH)_2$, can become thermodynamically favored over oligomers at circumneutral or higher pH.

Although there are numerous references to arsenite sulfide reactions in the literature, the influence of sulfide on the preservation and speciation of arsenic has not been systematically studied. Recently, Smieja and Wilkin 2003 reported that upon acidification to pH <2 with HCl or HNO_3, instaneous precipitation of poorly crystalline As_2S_3(am) occurred in sulfidic water, which resulted in the loss of dissolved arsenic. It is important to note that all their experiments were conducted at (0.35 to 10 mg/L) of As(III), i.e., concentrations that are much higher than those found in drinking water supplies. For the preservation of arsenic species in drinking water, EDTA-HAc, (Bednar et al. 2002) has been used, and for field speciation, acidification with H_2SO_4 (Edwards et al. 1998) has been widely used. However, the influence of S^{2-} on samples acidified with EDTA-HAc or H_2SO_4 has not been studied.

Clifford et al. 2004 addressed the influence of the presence of Fe(II) on arsenic speciation in simulated ground waters and in actual field samples. But the effect of S^{2-} on the speciation was not studied. Recently, Jay et al. (Jay et al. 2004) reported that at pH 6.0, charged As(III)-sulfide species would be formed in waters containing As(III) and sulfide, and would be effectively removed by anion-exchange during As(III)/(V) speciation. So, the speciation of arsenic in sulfidic waters will likely report thioarsenite species as As(V).

As_2S_3(am) will likely precipitate when pH is reduced by acidification and this will result in a loss of total arsenic. To determine the true arsenic concentration in the acidified sulfidic water, a three-step method involving base addition, oxidation, and re-acidification (Clarke and Helz 2000, Smieja and Wilkin 2003) has been proposed. H_2O_2 was used as an oxidant under alkaline conditions, because under acidic conditions H_2O_2 is not an effective oxidizer of arsenic sulfide. The reaction of As_2S_3(am) with H_2O_2 at pH > 11 to produce As(V) and sulfate is shown below:

$$As_2S_3(am) + 14\ H_2O_2 + 12\ KOH = 2K_3AsO_4 + 3\ K_2SO_4 + 20\ H_2O$$

After oxidation, the arsenate solution is acidified with HNO_3 ($pH<2$) to measure As(Tot) by hydride generation AAS. The H_2O_2 method is complicated, time consuming and uses hazardous H_2O_2. Thus, a simpler, less hazardous method would be preferred to measure As(Tot) in sulfidic waters.

In this work, the influences of sulfide on As(III)/(V) preservation and speciation, and on the determination of As(Tot) were studied in detail.

MATERIALS AND METHODS

Reagents

All reagents used were of analytical reagent grade. MilliQ water (18 MΩ) was used to prepare all solutions. Sulfide solution was prepared using washed crystals of $Na_2S.9H_2O$ in low DO MilliQ water. The oxidizing agents, H_2O_2, NaOCl, potassium peroxymonopersulfate (Oxone), peracetic acid were purchased from Sigma-Aldrich Co, MO.

Sulfide Analysis

Samples were analyzed for dissolved sulfides using a methylene blue method as described in the EPA method 376.3 and Standard Methods 4500-S^{2-} (APHA, AWWA, and WEF 1992). In the presence of N, N-dimethyl-p-phenylenediamine oxalate and ferric chloride, sulfide forms methylene blue, the absorbance of which was measured at a wavelength of 690 nm in a Hach spectrophotometer.

Preparation of As(III)-Sulfide Containing NSF 2 Challenge Water

For all laboratory experiments, sulfide-spiked NSF 2 Challenge Water (See chapter 2) was used. Sulfide in water gradually oxidizes in the presence of dissolved oxygen (Wilmot 1988). To prevent sulfide oxidation in the challenge water prior to the addition of the preservative, the water was slowly purged with N_2 for 1.5 h. The experimental set up was shown in Chapter 2 (See Figure 2.1). Challenge water of pH ~7.5 was prepared in a brown glass bottle of 2.7-L capacity. As(III) was spiked into the low-DO challenge water, and then freshly prepared S^{2-} solution was spiked under the sparging condition. To attain As-S equilibrium, the solution was kept for 5-10 minutes under N_2 at lower flow rate, ~ 300 mL/min. Fe(II) was also spiked into the challenge water when required after addition of S^{2-}. The first 500-mL (approx.) of the challenge water was wasted through the sample collection outlet tube, then the challenge water was collected with the required amount of preservatives or directly used for experiments.

Sample Preservation and Speciation

Samples were collected directly into the 250-mL pre-cleaned PP bottles, which contained 10.8 mL of 2M acetic acid and 3.35 mL of 0.1 M EDTA. A few samples were collected with 0.05% (v/v) H_2SO_4 and 0.15% (v/v) HNO_3. For speciation, both chloride- (Clifford et al. 2004, Le et al. 2000) and acetate-form (Edwards et al. 1998) anion-exchange resin columns were used.

Detailed descriptions of the speciation methods have been given in our previous AwwaRF report 2663 (Clifford et al. 2004) and in the papers of Edwards et al. 1998 and Le et al. 2000.

Sample Treatment for the Determination of As(Tot) in Sulfidic Water

Four different oxidizers, commonly used for oxidation of hydrogen sulfide in water and wastewaters were used for oxidative dissolution of amorphous arsenic sulfide. These oxidizers are hydrogen peroxide (H_2O_2), sodium hypochlorite (NaOCl), peracetic acid (CH_3COOOH), and potassium peroxymonopersulfate (Oxone). Initially, working solutions of the oxidizers were prepared with proper dilution with MilliQ water. Predetermined volumes of the prepared oxidizer solutions were dispensed into 5 mL of acidified (EDTA-HAc or HNO_3 or H_2SO_4) As-S solution. After a predetermined period of time, the oxidized sample was diluted to 10 mL and then the diluted sample was used to measure As(Tot) using the L-cystiene HGAAS technique. All experiments were run in triplicate.

Experiments with Cu(I) for Speciation

Attempts were made to make neutral water-soluble copper thioarsenite complexes using Cu(I) in sulfidic challenge water with and without EDTA-HAc. A solution of 4 mg/L Cu(I) solution was prepared in a very dilute HCl. Predetermined amounts of the Cu(I) solution were added to 250 mL preserved and unpreserved samples. After mixing, each sample was speciated in a minicolumn containing Cl-form anion-exchange resin.

RESULTS AND DISCUSSION

Effect of Sulfide on As(III)/(V) Speciation Without Preservatives

The literature survey revealed that under reducing conditions several As-S compounds may form depending on pH and the concentrations of arsenic and sulfide. When the chemical equilibrium modeling program MINEQL+ (Schecher and McAvoy 2001) is used with thermodynamic data from the recent literature the result is that at near-neutral and alkaline pH, when arsenite and sulfide are present together, they tend to form several soluble thioarsenite species. Preliminary experiments were conducted to observe the effect of S^{2-} on the stability of As(III) in an anoxic environment with time. As(III) was spiked into the low-DO-challenge water (pH 7.5 ± 0.1) so that the final concentration of As(III) was 50 µg/L. The Eh of the solution after addition of As(III) was 100 ± 10 mV. The purging was continued for an additional 4-5 minutes, then the N_2 flow was reduced to ~ 300 mL/min into maintain the anoxic environment. Under these conditions, duplicate samples were passed through a 0.2-µm syringe filter followed by a 7-mL anion-exchange resin column by a peristaltic pump at the flow rate ~ 24 mL per minute. The first 15 mL of the effluent was wasted and then 35 mL of the effluent was collected in a plastic tube. Triplicate speciated samples were collected using three different resin speciation columns. The schematic diagram of the experimental set up is shown in Figure 5.1. Samples were also collected without filtration or speciation for the determination of total arsenic. To the As(III)-containing challenge water with N_2 purging at low flow rate, a known amount of S^{2-} solution was spiked to maintain the S^{2-} concentration of ~2 mg/L. Approximately, five minutes were allowed for proper mixing of S^{2-} in the challenge water, then duplicate samples were collected directly in plastic tubes to determine As(Tot). For speciation, the sample was passed through the syringe filter (0.2 µm) before entering the resin column. After 1 h and 3 h, the sample was speciated as before. The results of the As(Tot) and As(III) determinations are given in Table 5.1. As(III) concentration in the low-DO challenge water without S^{2-} addition was 50.2 µg/L (Expts. 1a & 1b) and As(III) concentration in the speciated effluent was 49.5 µg/L (Experiments. 2a, 2b & 2c). When S^{2-} was spiked into the low-DO challenge water, the Eh was reduced to -300 mV. S^{2-} concentration at the beginning was 2.21 mg/L. Arsenic concentration was also measured before and after speciation.

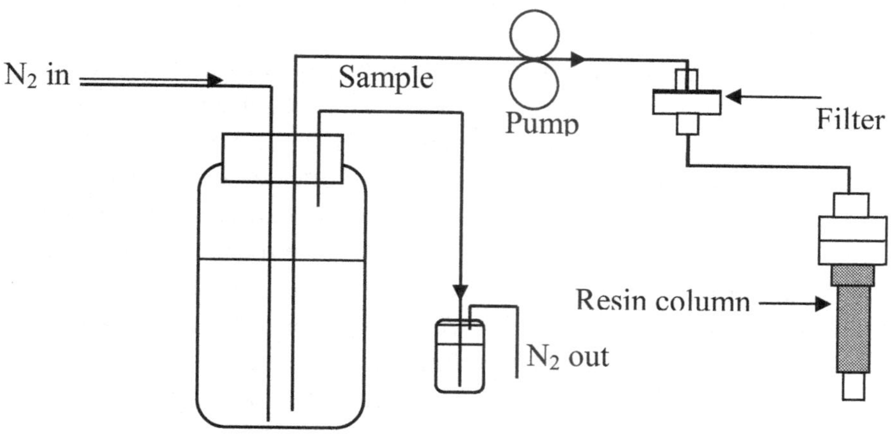

Figure 5.1 Flow schematic of the apparatus used to study the influence of sulfide on arsenic speciation.

Table 5.1
Influence of sulfide in As (III) determination

#	Conc of S^{2-}. (mg/L)	Filtration	Speciation	Time of collection.	As(tot) conc. (µg/L)	As(III) conc (µg/L)	Mean (µg/L)
1a	0	No	No	5 min after As addition	49.9	-	
1b	0	No	No	5 min after As addition	50.5	-	50.2
2a	0	Yes	Yes	5 min after As addition	-	49.3	
2b	0	Yes	Yes	5 min after As addition	-	49.3	
2c	0	Yes	Yes	5 min after As addition	-	49.8	49.5
3a	2.2	No	No	5 min after S^{2-} addition	48.5	-	
3b	2.2	No	No	5 min after S^{2-} addition	51.2	-	49.9
4a	2.2	Yes	Yes	5 min after S^{2-} addition	-	25.4	
4b	2.2	Yes	Yes	5 min after S^{2-} addition	-	23.6	
4c	2.2	Yes	Yes	5 min after S^{2-} addition	-	25.6	24.9
5a	2.2	Yes	Yes	1 h after S^{2-} addition	-	19.9	
5b	2.2	Yes	Yes	1 h after S^{2-} addition	-	18.6	
5c	2.2	Yes	Yes	1 h after S^{2-} addition	-	18.9	19.2
6a	2.2	Yes	Yes	3 h after S^{2-} addition	-	12.6	
6b	2.2	Yes	Yes	3 h after S^{2-} addition	-	12.7	
6c	2.2	Yes	Yes	3 h after S^{2-} addition	-	12.4	12.6

The concentration of As(Tot) in the feed water 5 min after the addition of S^{2-} was 49.9 µg/L, which was the same as before the addition of S^{2-}(Table 5.1). But, the concentration of As(III) in the speciated effluent was 24.9 µg/L i.e., about 50% lower than the feed and speciated sample before addition of S^{2-}. After 1 h, sulfide-containing challenge water was speciated again, and the result showed even lower As(III) concentration (19.2 µg/L) in the speciated effluent than before. Three hrs after the addition of S^{2-}, the sample was speciated again and the speciated As concentration (Expt 6a, 6b, & 6c) was reduced further to 12.6 µg/L.

From the above experimental results, it was concluded that when both As(III) and sulfide were present in anoxic water at circumneutral pH, negatively charged thioarsenite species were formed which were adsorbed onto the anion-exchange resin column. Jay et al. (Jay et al. 2004) also reported the adsorption of arsenic when As(III) containing sulfidic water at pH 6 was added to Bio-Rad acetate-form anion-exchange resin. They also reported that the full recovery of arsenic from the resins was not accomplished using 1N HCl for 10 min, although this elution regime was typically sufficient for As(V). According to Rader et al. (Rader et al. 2004), when the ratio of total sulfide to total As(III) is equal to or greater than two, virtually all As(III) is in the dithoarsenite form. Our experimental results (Table 5.1) indicate that with increasing time, the availability of undissociated H_3AsO_3 decreased. Helz et al. 1995 reported that in undersaturated solutions thermodynamically favored monomers, $AsS(OH)(SH)^-$ and $AsS(SH)_2^-$ were formed at circumneutral or higher pH. Wilkin et al. (Wilkin et al. 2003) experimentally identified and quantified various thioarsenic species by using IC-ICP-MS in sulfide concentrations > $10^{-4.3}$ M (1.6 mg/L) within the pH range 6-10. So, speciation of arsenic in sulfidic waters is a great challenge due to the formation of negatively charged thioarsenite species, which will be sorbed onto anion resins and interpreted incorrectly as As(V).

Effect of Sulfide in Speciation of As(III) with Preservatives

Separation of inorganic As(III)/(V) in the EDTA-HAc preserved samples was accomplished using two methods: differential pH-AAS and the anion-exchange resin column method (Clifford et al. 2004). Both methods were applied for speciation of arsenic in the presence of different amounts of sulfides. EDTA-HAc was used to preserve the arsenic species 10-15 minutes after addition of sulfide. The experimental results are shown in Figure 5.2, which indicates that at S^{2-} concentrations >0.5 mg/L, the differential pH-HG-AAS method gave lower concentrations of As(III) than the ion-exchange minicolumn field speciation method. This may be due to the formation of As-S species, which do not generate arsine at the experimental conditions.

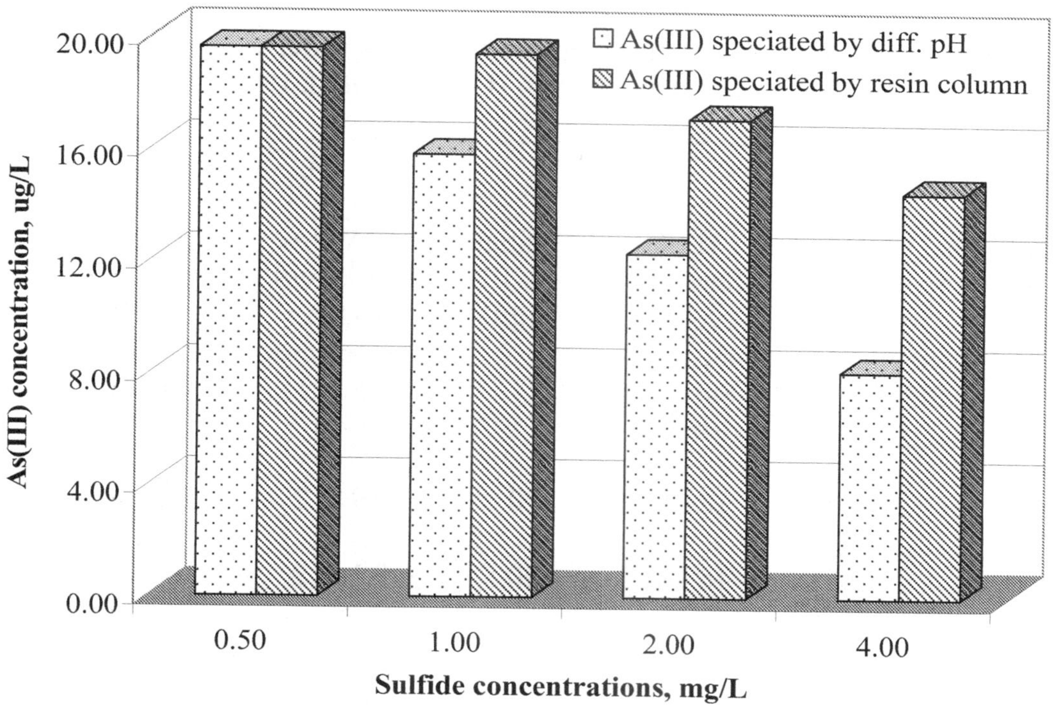

Figure 5.2. Comparison of As(III) concentrations measured by the differential pH method (citric acid/citrate buffer pH 5.0) and the Cl-form anion-exchange minicolumn method. Samples were preserved 15 minutes after sulfide addition and speciated within 5 minutes of preservation.

Among the many arsenic field speciation methods, the speciation of arsenic by the EDTA-HAc method at reduced pH 3.2 (Clifford et al. 2004) and the H_2SO_4 preservation speciation method at pH <2 (Edwards et al. 1998) have been well researched, but the effect of sulfide on these methods has not been studied. In the presence of sulfide, at low pH, the chemistry of As-S becomes more complicated with the formation of amorphous As_2S_3(am). Experiments were carried out to study the effect of sulfide on arsenic speciation of preserved samples at low pH. Low-DO-challenge water was spiked with 20 µg/L As(III) and 0.5 to 2 mg/L S^{2-}. Within 15 minutes of S^{2-} addition a sample of the spiked challenge water was collected and preserved with 1.34 mM EDTA + 8.7 mM HAc or 0.05%(v/v) H_2SO_4. After mixing, the preserved samples were immediately filtered and speciated by the Clifford et al. method and the Edwards et al. method. Experiments were done in triplicate on two different days. The results are shown in Table 5.2 and Figure 5.3, which indicate that up to 1 mg/L S^{2-}, in the presence of EDTA-HAc, the loss of the arsenic was not significant but pH 2.0 H_2SO_4 had severe effect on the arsenic speciation even at 0.5 mg/L S^{2-} concentration. The pH of the preserved sample played an important role in the influence of preservation speciation accuracy. The pH of the EDTA-HAc-preserved sample was ~3.2, whereas the pH of the H_2SO_4-preserved sample was <2.0. At pH 2.0, a significant portion of the As(III) precipitated as As_2S_3(am), whereas little if any precipitate formed at pH 3.2. Initially, in the presence of S^{2-}, at pH 7.5, under reducing condition, As(III) was present in the form of $H_2AsOS_2^-$/ $H_2AsS_3^{-1}$ and/or H_3AsO_3. When pH was

Table 5.2.
Ion-exchange speciation of As(III) in the S^{2-} spiked challenge water using the EDTA-HAc (Clifford et al. 2004) and H_2SO_4 (Edwards et al. 1998)[1]

Conc. of S^{2-} mg/L	Day 1			Day 2		
	Initial As(III) µg/L	EDTA-HAc (Speciation) As(III) conc.(mean± Std) µg/L (n=3)	H_2SO_4 (Speciation) As(III) conc.(mean± Std) µg/L (n=3)	Initial As(III) µg/L	EDTA-HAc (Speciation) As(III) conc.(mean± Std) µg/L (n=3)	H_2SO_4 (Speciation) As(III) conc.(mean± Std) µg/L (n=3)
0.5	20.6	20.1±0.4	18.6±0.4	20.1	19.9±0.2	18.3±0.2
1.0	20.5	19.5±0.5	16.7±1.3	20.5	19.1±0.2	16.7±0.6
2.0	NA	NA	NA	20.0	16.4±0.2	9.6±0.9

[1] Preservation carried out 15 min after S^{2-} spiking. Speciation carried out within five minutes of preservation. Experiments run on two different days.
NA = Not available

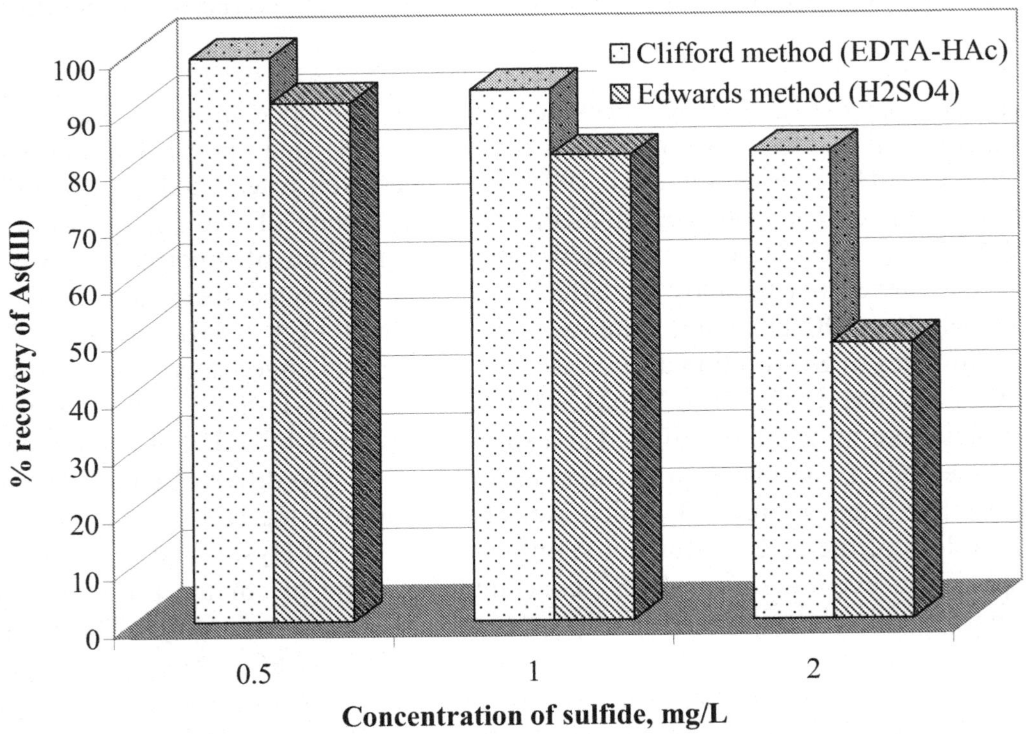

Figure 5.3. Comparison of IX-speciated As(III) concentrations as a function of sulfide concentration after preserving with EDTA-HAc or H_2SO_4. Cl-form resin was used for EDTA-HAc and Acetate-form resin was used for H_2SO_4.

reduced to 3.2 or 2.0, at least two different reactions (Equations 1 and 2, below) were possible depending on the pH, and the S^{2-} and As(III) concentrations.

$$H_2AsOS_2^- + H^+ + 2\,H_2O \leftrightarrow H_3AsO_3^o(aq) + 2\,H_2S(aq) \qquad (1)$$
$$2\,H_2AsOS_2^- + 2\,H^+ \leftrightarrow As_2S_3(am) + H_2S + 2\,H_2O \qquad (2)$$

When pH was low, e.g. pH 2.0 with H_2SO_4, As_2S_3 (am) precipitated according to Eq. (2). The mass of As_2S_3(am) that precipitated depended on the concentrations of S^{2-} and As(III).

To further elucidate the reactions of As(III) and S^{2-} in the presence of EDTA and HAc at pH 3.2, additional experiments were carried out at varying sulfide concentrations. Fifteen minutes after spiking with S^{2-}, the sulfidic water was preserved with EDTA-HAc. Both the preserved and unpreserved sulfidic waters were speciated using Cl-form anion-exchange resin. The results are shown in Figure 5.4, which indicates that the concentration of undissociated As(III) decreased with an increase in sulfide concentration in the unpreserved sample, probably due to formation of anionic thioarsenite species.

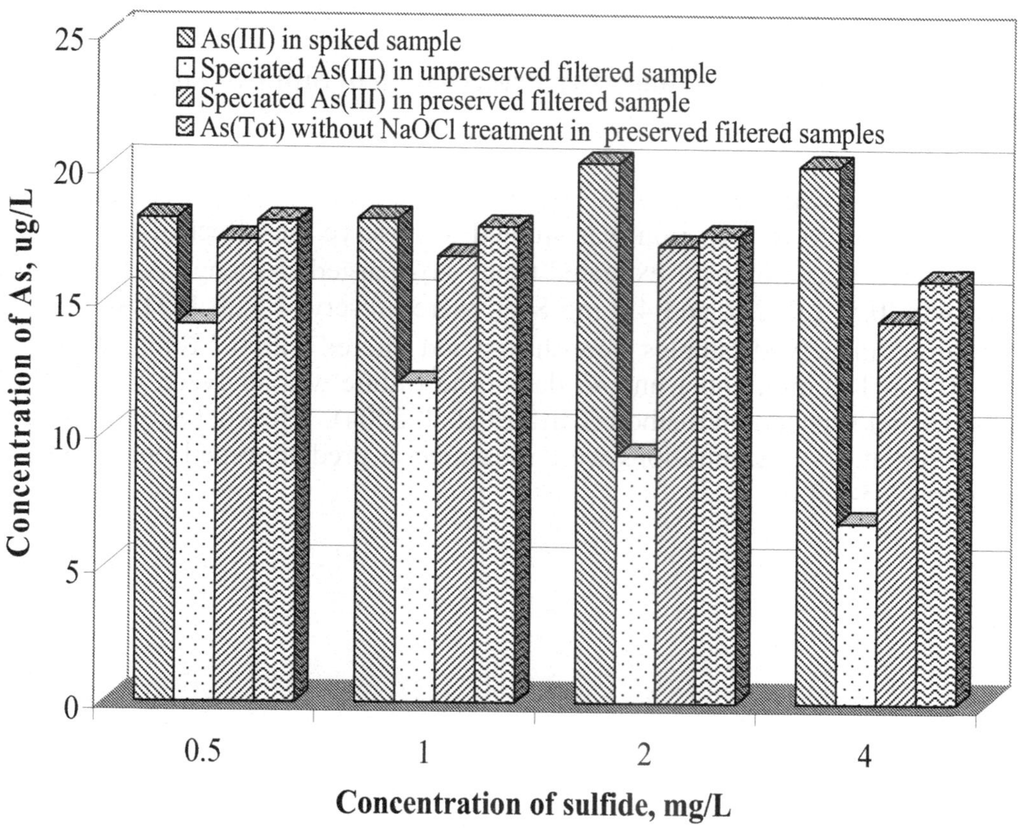

Figure 5.4. Comparison of As(III) and As(Tot) concentrations in the EDTA-HAc preserved and unpreserved samples as a function of sulfide concentration.

Interestingly, the concentrations of As(III) in the preserved samples were higher than the unpreserved samples. The high values of speciated As(III) in the EDTA-HAc preserved samples were due to shifting of the equilibrium of Eq (1) towards the right with the formation of uncomplexed, neutral arsenite. When As(Tot) concentrations were measured in the preserved samples (without oxidative dissolution of arsenic sulfide), the concentrations of As(Tot) were very close to the amount of As(III) spiked when sulfide was ≤ 1.0 mg/L which indicates that the formation of As_2S_3(am) up to 1 mg/L of sulfide was not significant. However, the As(Tot) concentration decreased significantly at sulfide concentration >1 mg/L due to the loss of arsenic by precipitation of As_2S_3(am). According to Rochette et al. (Rochette et al. 2000), at pH 4, arsenite is thermodynamically favored over diothioarsenite at H_2S < 87 μM (~2.8mg S^{2-}/L) and with the decrease of pH and increase of S^{2-}, As(III) was precipitated as As_2S_3(am). In the case of EDTA-HAc (pH 3.2) up to 1 mg/L of S^{2-}, the recovery of As(III) was ≥90%, but above 2 mg/L S^{2-}, significant loss of As(III) occurred due to the precipitation of As_2S_3(am). Based on the experimental results, it was concluded that in the presence of EDTA-HAc, up to 1 mg/L of S^{2-}, As(III) was more stable and not precipitated as As_2S_3(am). But when the concentration of S^{2-} was >1.0 mg/L, precipitate of As_2S_3(am) occurred according to the following reaction (Eq. 3).

$$2\ H_3AsO_3(aq) + H_2S(aq) \leftrightarrow As_2S_3(am) + 6\ H_2O \quad (3)$$

On the other hand, in the presence of H_2SO_4 even at 0.5 mg/L of S^{2-}, significant loss of As(III) took place due to the formation of As_2S_3(am) at pH <2.0 by the reaction (Eq. 2).

Changes in As(III) Speciation with Time after Preservation of Sulfide-Containing Samples Using EDTA-HAc

The solubility of As(III) in the EDTA-HAc preserved challenge water containing varying sulfide concentration and no ferrous iron was studied to observe the effect of the EDTA-HAc preservative on As(III) speciation over extended time. N_2-sparged challenge waters with 20 μg As(III)/L spiked with 0.5, 1.0, 2.0, and 4.0 mg S^{2-}/L were preserved with EDTA-HAc after 15 minutes equilibration. Preserved samples were shaken and filtered with 0.2 μm membrane filters at 2, 5, 10, 30, 60 and 180 min, and 1 and 45 days. The filtrate was analyzed for As (Tot) and As(III) using the IX field speciation method (Clifford et al. 2004). The concentrations of As(III) and As(Tot) at different S^{2-} concentrations were measured at predetermined times. The results are shown in Figures 5.5-5.8.

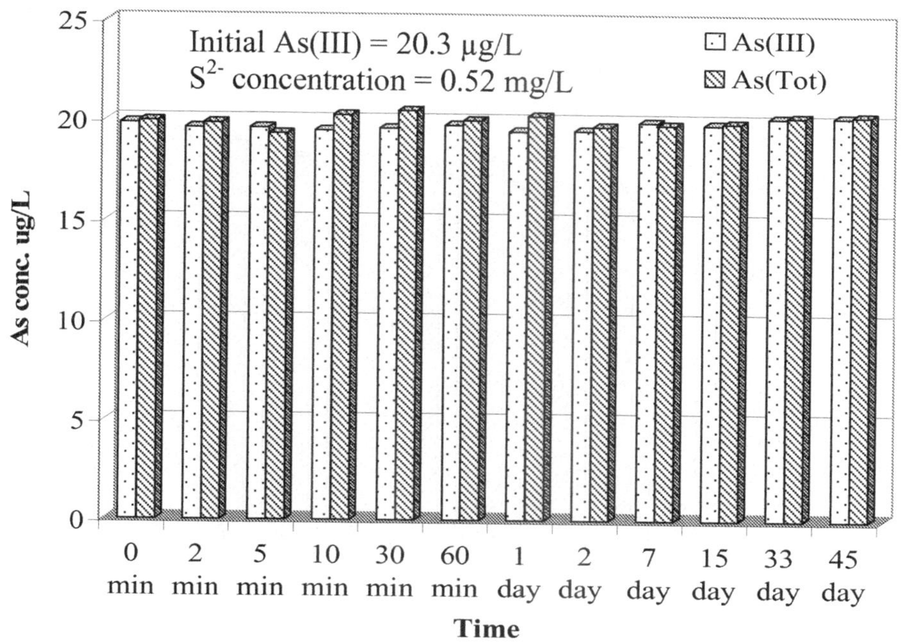

Figure 5.5. Concentrations of As(III) and As(Tot) measured at different time up to 45 days in the EDTA-HAc preserved challenge water with 0.5 mg/L sulfide. As(III) was speciated by anion-exchange resin column. Samples were preserved 15 min after sulfide addition and speciated at the indicated times.

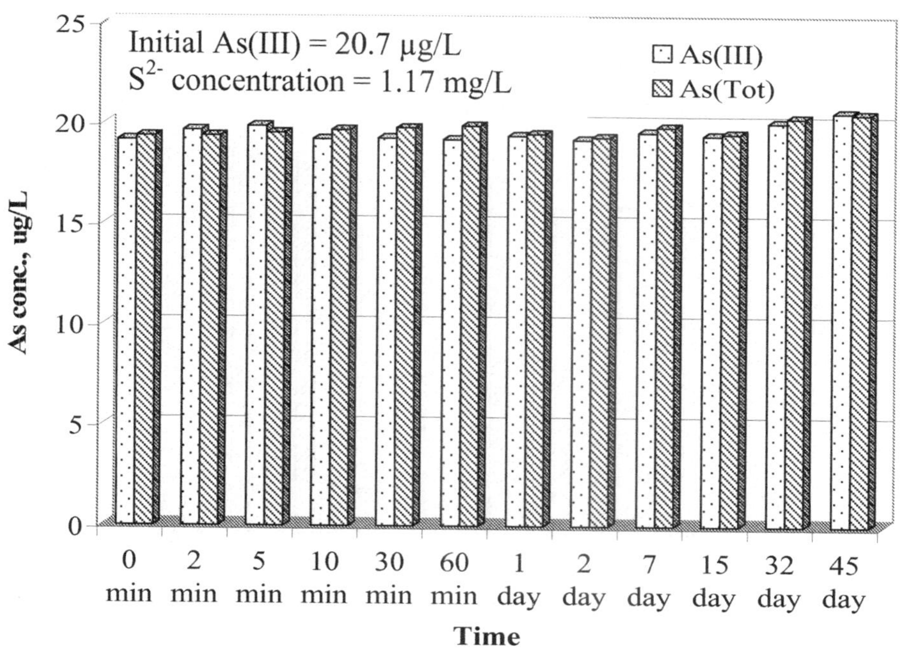

Figure 5.6. Concentrations of As(III) and As(Tot) measured at different time up to 45 days in the EDTA-HAc preserved challenge water with 1.0 mg/L sulfide. As(III) was speciated by anion-exchange resin column. Samples were preserved 15 min after sulfide addition and speciated at the indicated times.

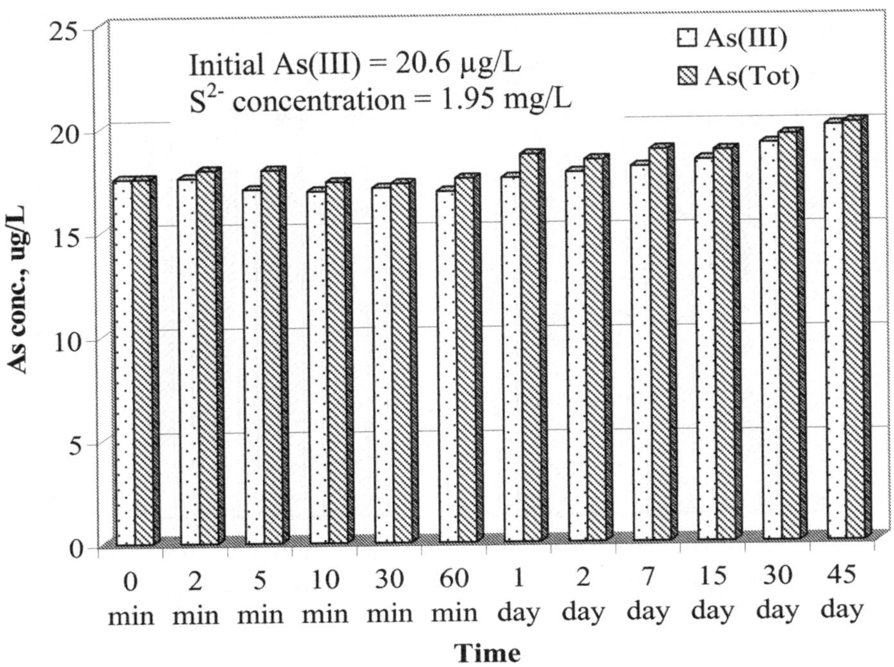

Figure 5.7. Concentrations of As(III) and As(Tot) measured at different time up to 45 days in the EDTA-HAc preserved challenge water with 2.0 mg/L sulfide. As(III) was speciated by anion-exchange resin column. Samples were preserved 15 min after sulfide addition and speciated at the indicated times.

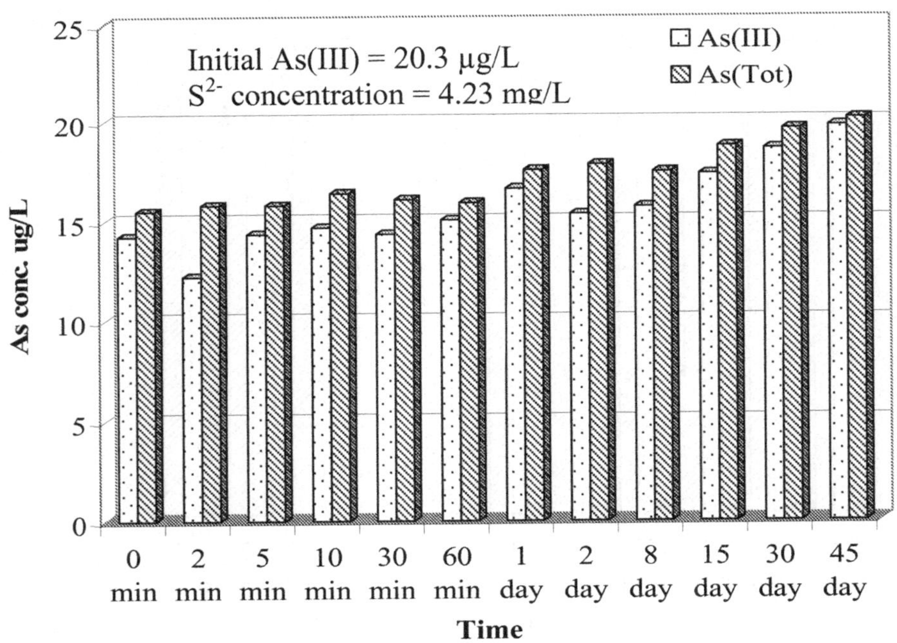

Figure 5.8. Concentrations of As(III) and As(Tot) measured at different time up to 45 days in the EDTA-HAc preserved challenge water with 4.0 mg/L sulfide. As(III) was speciated by anion-exchange resin column. Samples were preserved 15 min after sulfide addition and speciated at the indicated times.

From Figures 5.5-5.8 it was concluded that the solubility of As(III) in the EDTA-HAc preserved samples depended on the S^{2-} concentrations. Up to 1.0 mg/L of S^{2-} the effect was not so significant but at the higher concentrations of S^{2-}, the availability of As(III) was reduced due to the formation of As_2S_3(am) according to reaction (Eq. 3). Another interesting observation is that the concentrations of arsenite and As(Tot) in the preserved sample steadily increased during the preservation period of 45 days. This was more pronounced when S^{2-} concentration was >1.0 mg/L (Figures 5.7 and 5.8). The increase of arsenite in the preserved sample with time was governed by two factors: (1) slow dissociation of thioarsenite species with time, and (2) spontaneous solubilization of As_2S_3(am) with time by as shown in Eqs. 4 and 5, below. The solubility of As_2S_3(am) and the formation of arsenic species depended on pH and concentration of S^{2-}. Floroiu et al. (Floroiu et al. 2004) in their kinetics study of dissolution of As_2S_3 within the pH range 2 to 8 reported that the formation of negatively charged thioarsenite species dominated over arsenite at near neutral to alkaline pH when sulfide concentrations were greater than about 0.1 to 1 mM (3.2 to 32 mg/L). The following primary dissolution reactions have been suggested to control the solubility of As_2S_3 (Eary 1992, Helz et al. 1995, Clarke and Helz 2000):

$$0.5\ As_2S_3(am) + 3\ H_2O \leftrightarrow H_3AsO_3 + 1.5\ H_2S \quad (4)$$
$$0.5\ As_2S_3(am) + H_2O + 0.5\ H_2S \leftrightarrow AsS(OH)(SH)^- + H^+ \quad (5)$$

In acidic, sulfide-deficient solutions, the first reaction (Eq. 4) dominates, whereas in acidic solutions with excess sulfide, the second reaction (Eq. 5) plays a greater role. According to Clarke and Helz's model (Clarke and Helz 2000), at pH 2, the loss of As(III) from solutions by precipitation of As_2S_3 is not expected to occur with solutions containing 0.35 mg/L As(III) until the concentration of sulfide exceeds 0.8 mg/L. Again, with solutions containing 2.0 mg/L As(III), precipitation of As_2S_3(am) and loss of dissolved arsenic were predicted to occur when the sulfide concentration exceeded 0.25 mg/L. Similar experimental results were reported by Smieja and Wilkin (Smieja and Wilkin 2003) by acidifying the sulfidic arsenic solutions with HNO_3 and HCl at pH 2.0. In our present experimental conditions, 0.02 mg/L As(III) was present at pH 3.2. So the loss of As(III) might be expected only at relatively high sulfide concentrations. In our experiments with EDTA-HAc at pH 3.2 we did not find any significant loss of As(III) with sulfide concentration up to 1.0 mg/L. But with sulfide concentrations >1.0 mg/L a significant loss of As(III) occurred due to precipitation of amorphous As_2S_3, which spontaneously dissolved with time with the formation of H_3AsO_3 (Eq 4). Figures 5.7 and 5.8 clearly indicate the increase of As(III) in the preserved samples with time, and after 45 days, the concentrations of the dissolved As(III) were almost same as the initial concentrations of As(III) spiked.

Changes in As(III) Speciation with Time after Sulfide Addition and Prior to Adding the EDTA-HAc Preservative

In our previous experimental results, it was observed that within 15 minutes of spiking of $\leq$1.0 mg/L S^{2-}, the addition of EDTA-HAc increased the concentrations of the undissociated arsenite through the disproportionation of thioarsenite species by the reaction (Eq. 1) mentioned above. However, the kinetics of this reaction was not studied. Thus, the following experiment was carried out to determine the effect of the As(III)-S^{2-} reaction time on the As(III) preservation and speciation processes with EDTA-HAc at pH 3.2. In the low-DO-Challenge water, a known amount of As(III) was spiked under the N_2 sparging condition, and then approximately 1 mg/L

S^{2-} was spiked. The N_2 sparging was continued for another 2-3 minutes, and then the flow rate was reduced to ~300 mL/min to maintain the N_2 atmosphere in the bottle. The As-S-containing-challenge water was collected at predetermined times in a 100-mL volumetric flask with the required amount of EDTA-HAc or without any preservative. Both the preserved and unpreserved samples were filtered and speciated by a 7-mL Cl-form anion-exchange resin column. The effluents were analyzed for As(III) representing total arsenic after the treatment with NaOCl to re-dissolve any As_2S_3(am). The experimental results are shown in Figure 5.9, which indicates that in the presence of 0.89 mg/L S^{2-}, about 30% of the initial As(III) was converted to thioarsenite species within 15 minutes of the S^{2-} spiking. When EDTA-HAc

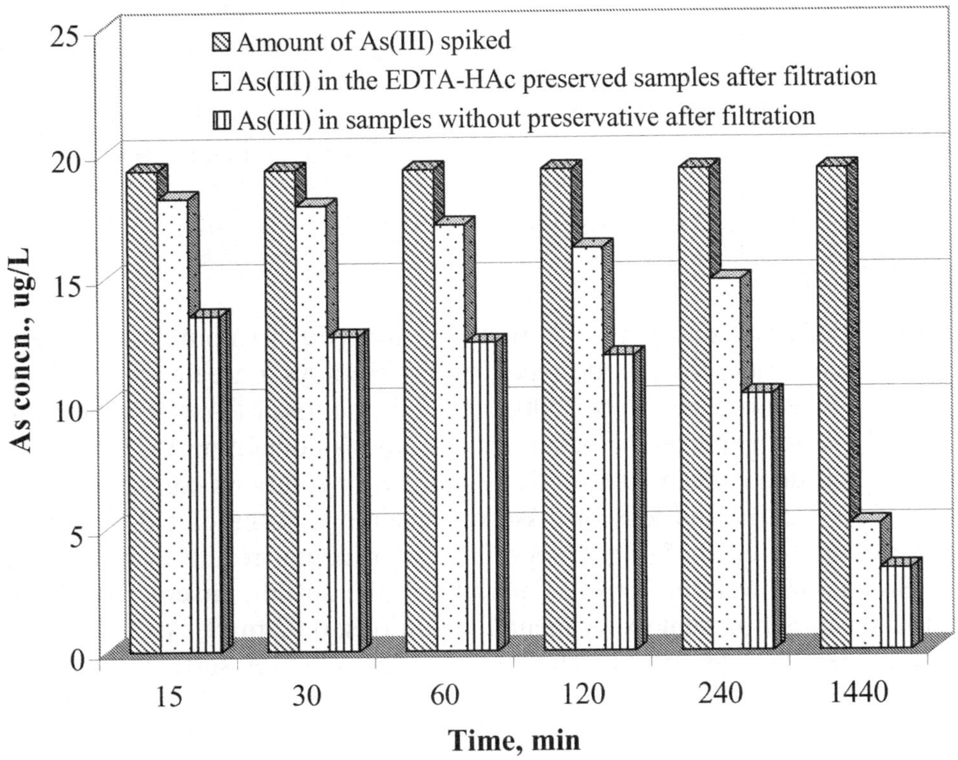

Figure 5.9. Soluble As(III) in preserved and unpreserved samples as a function of As(III)-S^{2-} equilibration time prior to preservation and ion-exchange speciation with EDTA-HAc (The S^{2-} concentrations were 0.89 mg/L initially, and 0.56 mg/L after 24 hr.)

was used for preservation of the sample equilibrated with arsenite and sulfide for 15 min, the concentration of arsenite increased and the recovery was about 94%, which was about 30% higher than the concentration of arsenite in the unpreserved sample. After 4 h of As(III)-S^2 equilibration, about 53% of the initial As(III) was converted to thioarsenite species but the addition of EDTA-HAc produced nearly 77% arsenite, which was 44% higher than the concentration of As(III) in the unpreserved sample. After 24 h of As(III)-S^2 equilibration, more than 80% of As(III) was converted to thioarsenite species and after preservation and immediate speciation, the recovery of As(III) was only 26% of the initial As(III). After 24 h, the S^{2-} concentration was reduced from 0.89 mg/L to 0.56 mg/L due to N_2-sparging losses. The experimental results indicated that under reducing conditions in the presence of about 0.5 to 1 mg/L S^{2-}, the arsenite originally present was 80% converted to negatively charged thioarsenite species within 24 hr. Therefore, in the typical sampling situation, groundwaters containing sulfide that has been in equilibrium with arsenite for a long time at near neutral pH, negatively charged thioarsenite species should be predominant, and the immediate speciation after addition of EDTA-HAc will not measure the true As(III) concentration in the water sample.

Speciation and Preservation of Arsenic in Sulfidic waters in the Presence of Iron

The concentrations of dissolved sulfide under anoxic conditions are controlled by the abundance of Fe(II) in the water. Although we did not expect Fe(II) to be soluble in sulfidic waters, we ran a few experiments to verify the expected influence of Fe(II) on arsenic speciation in the presence of sulfide. In one experiments, Fe(II) was spiked into 20 μg/L As(III) and 2 mg/L sulfide-containing challenge water at pH 7.5 under N_2 pressure at room temperature. Immediate formation of a black precipitate, presumably of Fe(II) monosulfide, took place after spiking Fe(II). This observation was in agreement with the expectation that in the natural water systems, dissolved sulfide concentrations are able to increase only after the supply of reactive iron is exhausted via iron monosulfide precipitation (Canfield 1989). It is expected that high-sulfide water will contain very low dissolved iron. In other words, the presence of reactive iron will reduce or eliminate the formation of thioarsenite species. The fate of arsenic in anoxic environments in the presence of iron monosulfides or iron disulfides is incompletely understood. At low sulfide concentrations complete adsorption of arsenic may take place. On the other hand, in the presence of high sulfide concentrations little absorption of arsenic occurs on the iron sulfide (O'Day et al. 2004). The lack of sorption may be due to the presence of thioarsenite species. Due to the immediate precipitation of iron sulfide in the screening experiment, no systematic study for speciation and preservation of arsenic in the presence of Fe and sulfide was attempted in this research.

Determination of Total Arsenic (As(Tot)) in Sulfidic Water

As a result of the experiments discussed earlier, it was concluded that As_2S_3(am) was precipitated from arsenic-containing sulfidic waters when preserved with acid. HNO_3, H_2SO_4, or a mixture of EDTA and HAc, i.e., preservatives that have commonly been used to preserve arsenic samples for immediate field speciation or days-later speciation. Therefore, the arsenic concentrations measured in the preserved samples could have been less than the true concentrations due to precipitation of As_2S_3(am) at low pH. The method used to measure As(Tot) in acidified sulfidic samples is oxidative dissolution of alkaline solution of As_2S_3(am) (pH >10.0) with H_2O_2. After oxidation, the sample is acidified with HNO_3 to measure arsenic concentration. When the pH is <10 during H_2O_2 addition, incomplete oxidation is observed (Smieja and Wilkin 2003). The problems when using H_2O_2 are as follows: (1) It is a time-consuming process. (2) Extra precautions are needed to reduce the loss of arsenic when raising the pH of the acid preserved sample and again acidifying the alkaline solution, and (3) hazardous chemicals are required. In the present work, attempts were made to oxidatively dissolve the As_2S_3(am) in the preserved samples using oxidants, which were expected to react with As_2S_3(am) in the acid medium and are less hazardous than H_2O_2.

For the oxidative dissolution study, 0.0027 M NaOCl (1 mL of 1.6×10^{-2} M NaOCl solution in 5 mL sample), potassium peroxymonopersulfate (Oxone) (1 mL of 0.1M oxone in 5 mL sample), peracetic acid (1 mL of 1% peracetic acid solution in 5 mL sample), and H_2O_2 (1 mL of 10 fold-diluted H_2O_2 (30%)) were used. Samples containing the oxidizers were held for 6 h before analysis of As(Tot). The experimental results are given in Table 5.3, where it can be seen that potassium peroxymonopersulfate (Oxone), peracetic acid, and H_2O_2 were not effective oxidizers in acid-preserved sulfidic water samples. Even the concentrated solutions of potassium peroxymonopersulfate (Oxone), H_2O_2, and peracetic acid did not improve the recovery of arsenic. The concentrated solution of potassium peroxymonopersulfate (Oxone) (1 M) or peracetic acid (2%) oxidized L-cysteine in the pre-reduction step of arsenic analysis. On the other hand, sodium hypochlorite gave satisfactory results for the recovery of arsenic even in the presence of 4 mg/L of S^{2-}. NaOCl is a strong oxidizer that dissociates into Na^+ and OCl^- ions. The hypochlorite ions are readily protonated to form HOCl, hypochlorous acid, a weak acid in acidic solution:

$$NaOCl \leftrightarrow Na^+ + OCl^- \qquad (6)$$
$$H^+ + OCl^- = HOCl \qquad (7)$$

In acidic solution, hypochlorous acid reacts with As_2S_3(am), $H_2AsOS_2^{-1}$, H_3AsO_3, and unreacted H_2S according to the follow reactions:

$$As_2S_3(am) + 6\ HOCl \leftrightarrow 2\ H_3AsO_3 + 3\ SCl_2 \qquad (8)$$
$$H_2S + HOCl \leftrightarrow S^o + H_2O + HCl \qquad (9)$$
$$H_2AsOS_2^- + 4\ HOCl^- + H^+ \leftrightarrow H_3AsO_3 + 2\ SCl_2 + 2\ H_2O \qquad (10)$$
$$H_3AsO_3 + HOCl \leftrightarrow H_3AsO_4 + HCl \qquad (11)$$

Based on the stoichiometry of reactions (8)-(11), 0.8 μM/L HOCl was required to oxidize the maximum possible amount of As_2S_3 present (20 μg As/L or 0.133 μM/L As_2S_3) while 0.125 mM/L HOCl was required to oxidize the maximum H_2S concentration present (4 mg/L or 0.125 millimolar H_2S). A small additional amount, 0.267 μM/L, of HOCl was required to oxidize maximum possible amount of H_3AsO_3 present. All the reactions were considered complete because a large excess, 2.67 mM/L, of HOCl was present.

Table 5.3.
Comparison of oxidizing efficiency of different oxidizers in acid-preserved arsenic-spiked unfiltered challenge water containing 3.86 mg/L S^{2-} at room temperature

As conc µg/L	Conc. of As after preservation, (µg/L) (without treatment)			Conc. of As after treatment with NaOCl (µg/L)			Conc. of As after treatment with Oxone (µg/L)			Conc. of As after treatment with peracetic acid (µg/L)			Conc. of As after treatment with H_2O_2 (µg/L)		
	EDTA -HAc (n=3)	HNO_3 (n=3)	H_2SO_4 (n=3)	EDTA -HAc (n=3)	HNO_3 (n=3)	H_2SO_4 (n=3)	EDTA -HAc (n=3)	HNO_3 (n=3)	H_2SO_4 (n=3)	EDTA -HAc (n=3)	HNO_3 (n=3)	H_2SO_4 (n=3)	EDTA -HAc (n=3)	HNO_3 (n=3)	H_2SO_4 (n=3)
22.2 ±0.1	18.0 ±0.3	17.8 ±0.2	17.6 ±0.3	22.0 ±0.3	22.1 ±0.2	21.2 ±0.3	18.9 ±0.5	19.5 ±0.5	18.1 ±0.2	19.4 ±0.2	18.5 ±0.1	18.7 ±0.2	19.6 ±0.2	19.4 ±0.3	18.9 ±0.3
19.9 ±0.2	16.4 ±0.1	16.1 ±0.2	N.A	19.8 ±0.2	19.7 ±0.3	N.A	18.1 0.3	18.5 ±0.1	N.A	18.0 ±0.4	17.9 ±0.3	N.A	18.7 ±0.3	18.4 ±0.5	N.A

N.A = Not available

The kinetics of the HOCl-sulfide reaction was also studied in an effort to establish an appropriate contact time to solubilize the As_2S_3(am) after addition of NaOCl. The experimental results are given in Figure 5.10, which indicates that oxidation of As-S solution by HOCl was very fast. Immediately after the addition of NaOCl, recovery of As was 98%. Chemical equilibrium was attained within 5 minutes after the addition of NaOCl solution. So, the advantages of using NaOCl are (1) no need to make alkaline the acid-preserved sample before addition of oxidizer, (2) no need to acidify after oxidation, (3) rapid oxidation (almost instantaneous) in comparison with H_2O_2, and (4) less hazardous to handle than H_2O_2.

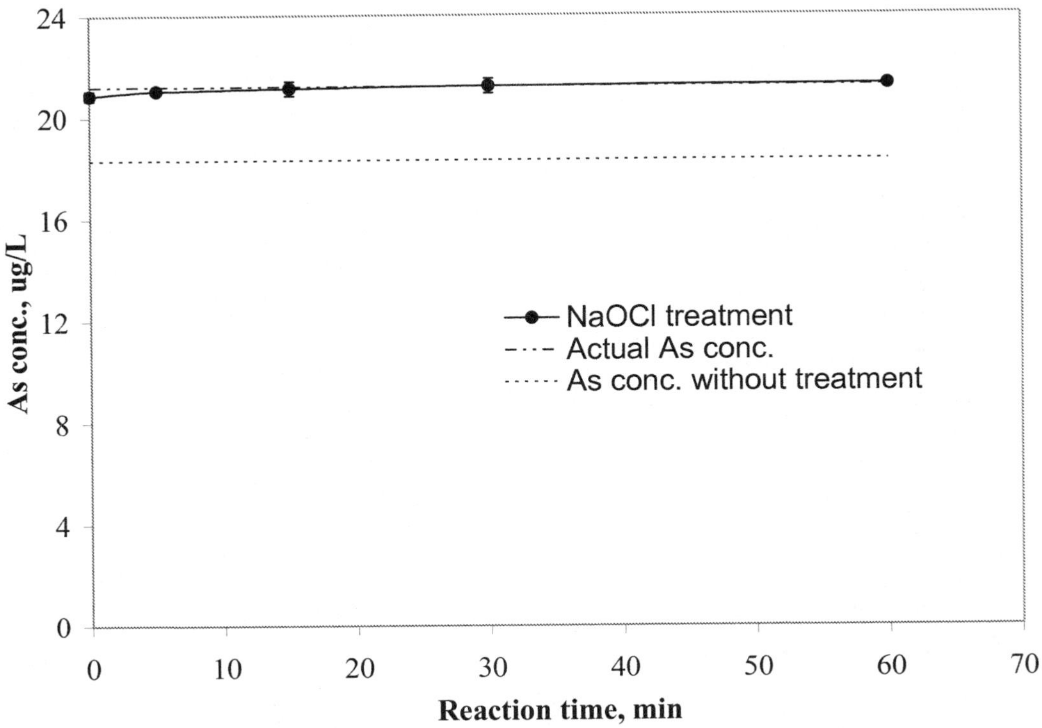

Figure 5.10. Total arsenic concentration (As(Tot)) as a function of NaOCl reaction time in an EDTA-HAc preserved As-S sample containing 21.2 μg/L As and 3.86 mg/L S^{2-} using 0.00267 M NaOCl at room temperature.

Speciation of As(III) and Thioarsenite Species

The recent literature shows that the bioavailability and toxicity of As(III) and thioarsenite species vary significantly (Tossell 2000 and Rader et al. 2004). So, it is important to know the actual concentrations of As(III) and thioarsenite species in water samples. Anion-exchange resin can be used for efficient separation of uncharged As(III) i.e., H_3AsO_3, and negatively charged thioarsenite species. At the circumneutral pH, under reducing conditions, and in the presence of sulfide, negatively thioarsenite species are formed, which will be adsorbed onto the anion-exchange resin when the water is passed through the ion-exchange column. The uncharged As(III) will pass through the column without retention. Laboratory experiments were carried out spiking known amounts of As(III) into sulfidic challenge water at pH 7.5. As-S-containing challenge water was kept under N_2 pressure to maintain the reducing environment. After 15 minutes, As-S-containing challenge water was forced through a 7-mL Cl- form anion-exchange resin column using a 60-mL syringe. The initial 15 mL displacement volume of the effluent was discarded, and the remaining 45 mL was collected in a 50-mL centrifuge tube with EDTA-HAc as a preservative. The original sample was also preserved with EDTA-HAc to measure As(Tot). The speciated and preserved As(Tot) samples were analyzed for As(Tot) after treatment with NaOCl to dissolve any As_2S_3(am) present. The concentration of charged thioarsenite species sorbed onto the resin was calculated from the difference of As(Tot) and the concentration of As(III) as H_3AsO_3 in the IX minicolumn effluent. The experimental results are shown in Figure 5.11, which indicates that the amount of arsenic retained by the resin column increased as the sulfide concentration increased. Correspondingly, the concentrations of As(III) as H_3AsO_3 decreased with an increase in sulfide concentration. It is also noted that as reported in Table 5.1 and as shown in Figure 5.9, increasing contact time between As(III) and sulfide produces more thioarsenite species.

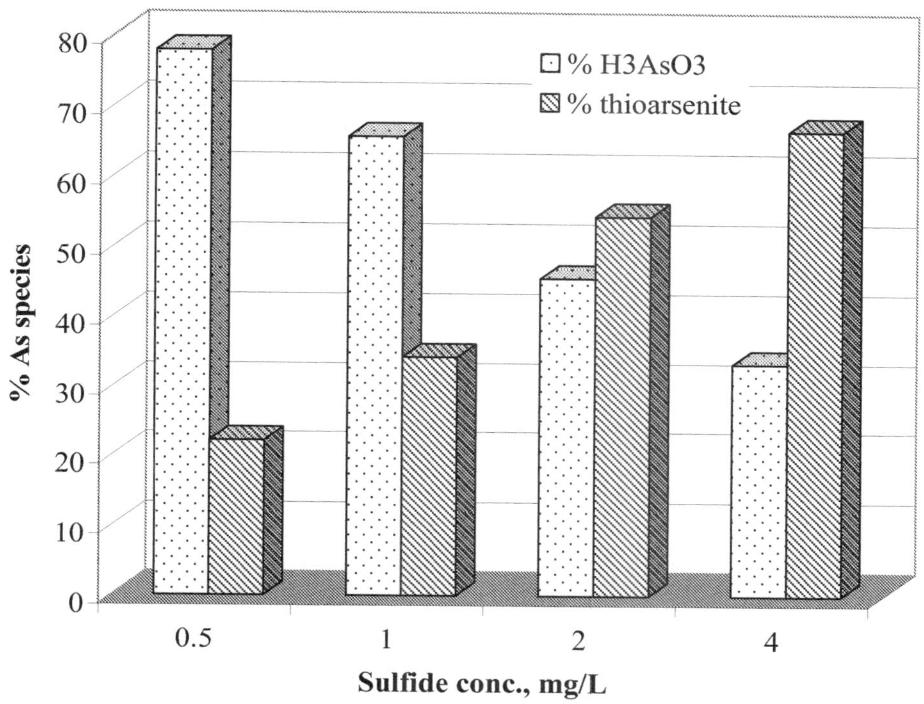

Figure 5.11. Concentrations of As(III) as H_2AsO_3 and thioarsenite species in the challenge water containing 20.5 µg/L of As(III) 15 min after adding different amounts of sulfide. Samples were speciated by Cl-form anion-exchange column.

According to Jay et al. 2004, the thioarsenite species adsorb very strongly onto the anion-exchange resin and are very difficult to elute. In their experiments, the use of a high concentration of acid (6 N HCl) eluent and prolonged contact time (7 h) did not recover 100% of the adsorbed arsenic due to the formation of orpiment (As_2S_3(am)). (In retrospect, strong acid is not reasonable to elute a sulfide species that precipitates in the presence of strong acid. A better approach would have been to elute with concentrated NaCl at neutral pH.) They recommended at least 24-h purging by pure N_2 before speciation to convert the thioarsenite species to arsenite.

In our present work, attempts were made to elute the adsorbed thioarsenite species from the anion-exchange resin column. In our previous experiments, it was observed that arsenic sulfide and/or thioarsenite species decomposed and oxidized very quickly with NaOCl solution with the formation anionic As(V) in the form of $HAsO_4^{2-}$ (Eqs. 8, 10, and 11). Thus, NaOCl solution was used to redissolve any As_2S_3(am) present and to oxidize thioarsenite species and elute As(V) from the anion-exchange resin column. Thirty mL of 20 µg/L As(III) in the presence of 3.75 mg/L S^{2-} was passed through a 7-mL chloride-form resin column after 15 to 20 minutes of addition of S^{2-} under reduced nitrogen pressure. The first 15 mL of the eluent was discarded and the remaining 15 mL was collected in a centrifuge tube and then the column was washed with 10 mL of MilliQ water. The concentrations of As(Tot) in the sulfidic water and in the speciated effluent were measured after the treatment with NaOCl. The initial As(III) as H_2AsO_3 concentration in the sulfidic water was 19.7 ± 0.3 µg/L (n=3) and the concentration of As(III) as H_2AsO_3 in the speciated effluent was 7.7 ± 0.2 µg/L (n = 6). So, under the experimental conditions, nearly 12 µg/L of thioarsenite species was formed, which was subsequently adsorbed onto the column. The adsorbed thioarsenite species was oxidized to As(V) and eluted with 0.16 M NaOCl solution. The elution profile is shown in Figure 5.12, which indicates that nearly 92% of thioarsenite was eluted from the column in the first 60 mL of the eluent and more than 99% was eluted in 100 mL of the eluent. So, complete recovery of thioarsenite species was possible by using NaOCl solution alone. (An alternative approach to eluting the thioarsenite species might be a mixture of dilute NaOCl to oxidize thioarsenite to As(V) and 1 M NaCl to elute the As(V)). The blank consisted of three 7-mL resin columns which were eluted with 0.16 M NaOCl solution. The eluent samples from the blank contained As(Tot) below the detection limit.

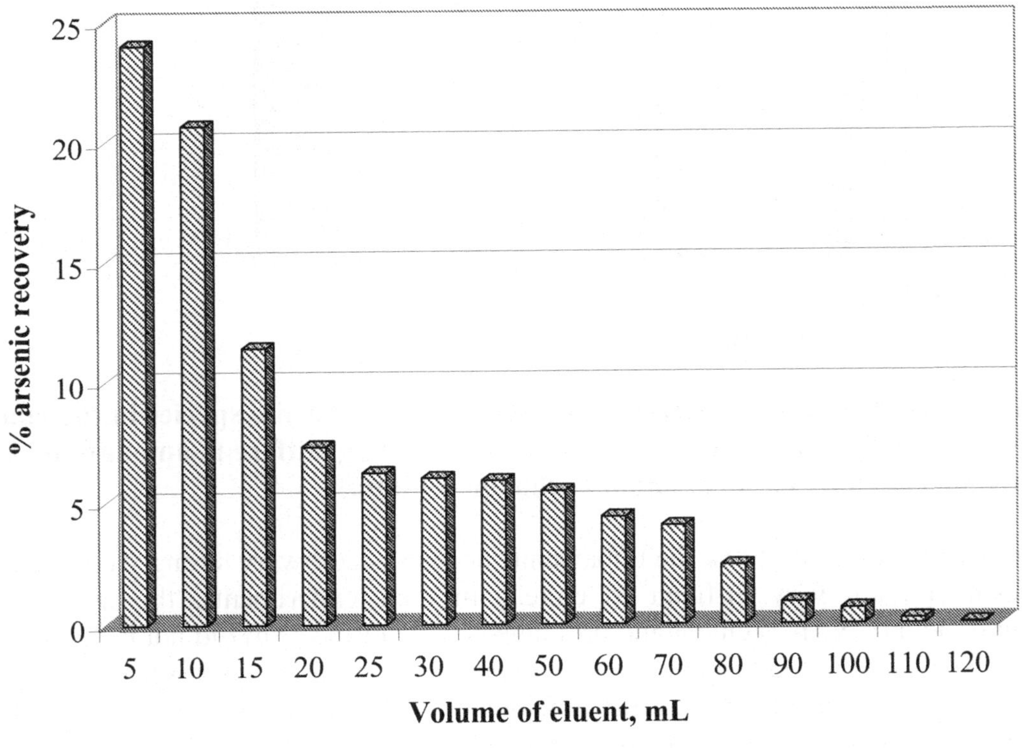

Figure 5.12. Recovery of thioarsenite species from anion exchange resin as a function of volume of 0.16 M NaOCl solution used for elution.

After finding that NaOCl solution alone could be used to oxidize and elute thioarsenite species from ion-exchange resin after speciation, the NaOCl concentration was optimized. Figure 5.13 shows that in the first 20 mL of the eluent maximum recovery was obtained by using 0.16 M NaOCl. At 0.32 M NaOCl, the recovery was lower than 0.16 M NaOCl solution possibly because of oxidation of the resin or the high pH during reduction of As(V) using L-cysteine. It is important to note that very dilute, 0.016 M, NaOCl solution did not elute arsenic from the resin column. So, there would be no loss of arsenic during washing of the thioarsenite loaded column with 10 mL MIlliQ water. This observation is consistent with the interpretation that arsenate ions produced from thioarsenite species by hypochlorite oxidation, are held on the ion exchange sites by electrostatic attraction and cannot be rinsed off with pure water only.

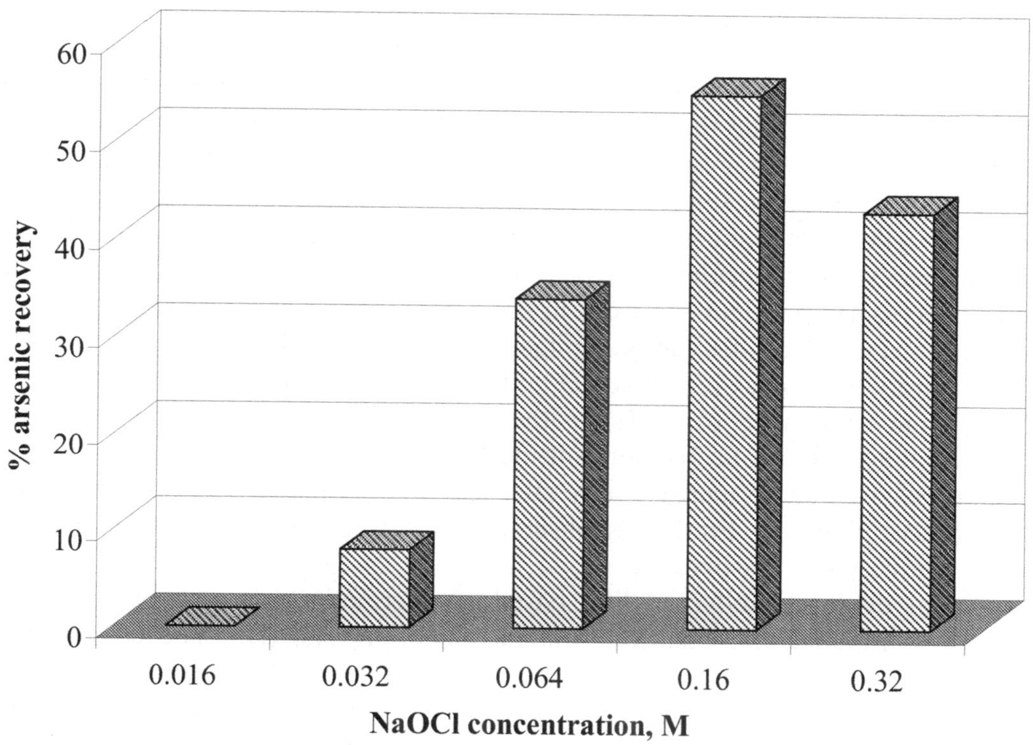

Figure 5.13. Recovery of thioarsenite species in the first 20 mL of eluent during elution with increasing concentrations of NaOCl.

To reduce the volume of the eluent required for thioarsenite elution, 3-mL Cl-form resin columns from Supelco instead of 7-mL columns were used for speciation of As(III) and thioarsenite species. Ten mL of As-S containing sample was passed through the 3-mL column after preconditioning. After discarding 4 mL of the effluent, the remaining 6 mL was collected for As(III) measurement. For elution of thioarsenite species from the column, 20 mL of 0.16 M NaOCl was used. The recovery of arsenic was 101% when nearly 20 μg/L of As(III) and 4 mg/L of sulfide were used for the preparation of test water.

Based on experimental and theoretical evidence, Clarke and Helz (Clarke and Helz 2000) and Tossell 2000 reported that Cu(I) forms stable aqua complex with thioarsenite ($Cu^+ + H_2AsOS_2^- \leftrightarrow CuH_2AsOS_2^{\circ}$, log K = 19.82 ± 0.17). Attempts were made to form Cu(I) complexes with thioarsenite species, but this approach was not successful. Most probably, Cu(I) thioarsenite complexes are formed in anaerobic environments. Under aerobic conditions, Cu(I) oxidizes to Cu(II), which does not form any complex with thioarsenite. Based on the preliminary results no further study was done in the laboratory.

SUMMARY, CONCLUSIONS, AND RECOMMENDATIONS*

Summary and Conclusions

Methods based on the addition of EDTA and acetic acid to a water sample were developed at the University of Houston for immediate field speciation (Project No. 2663, Clifford et al. 2004) and/or 30-day preservation (Chapters 1-4 of this report) of inorganic As(III)/(V) species. The same EDTA-HAc preserved sample with or without Fe(II) present could be held for 30 days and/or speciated in the field using Cl-form anion-exchange. Until now the effect of sulfide in the arsenic-contaminated water had not been considered. A literature search revealed that As(III) is the thermodynamically stable arsenic oxidation state present in reducing waters containing sulfide at neutral to alkaline pH. The As(III) can be present as H_3AsO_3 or several different thioarsenite species. As(V) is not thermodynamically stable in the presence of sulfide under reducing conditions. Thus, with waters containing arsenic and sulfide together, it is not necessary to speciate As(III)/V because As(V) cannot be present. Although speciation is not necessary, problems do exist in the determination of total arsenic (As(Tot)) in these waters because arsenic sulfides can precipitate in acid-preserved samples.

The concentrations of dissolved arsenite and thioarsenite species under anoxic condition are controlled by the abundance of Fe(II) in the water. In the presence of sulfide, Fe(II) precipitates immediately as black Fe(II) monosulfide. Thus, in the natural water system, dissolve sulfide concentrations are able to increase only after the supply of reactive iron is exhausted via iron monosulfide precipitation. From this observation, it is expected that sulfide-containing water should contain very low dissolved iron.

Experimental evidence and modeling based on thermodynamic equilibrium show the presence of thioarsenite species in sulfidic waters containing As(III). The concentration of thioarsenite species depends on the pH, and concentrations of S^{2-} and As(III). The objective of the research was to determine the effect of sulfide on the newly developed EDTA-HAc preservation-speciation method. The following conclusions could be made from the study.

1. In the presence of sulfide in at near neutral pH, As(III) was mainly present as thioarsenite species. The addition of EDTA-HAc to the sulfidic water lowered the pH to 3.2 and either increased the H_3AsO_3 concentration or precipitated amorphous arsenic sulfide, As_2S_3(am), depending on the sulfide concentration. Under this condition, As(III)/(V) speciation was not possible, and only total arsenic could be measured.
2. EDTA-HAc preserved samples with higher concentrations of S^{2-} showed immediate loss of total arsenic, but with time, the concentrations of As(Tot) and As(III) increased due to re-formation of H_3AsO_3 from the thioarsenite species and dissolution of As_2S_3(am). After 45 days, the recoveries of As(Tot) and As(III) were nearly 100%.

Note: This Summary, Conclusions, and Recommendations applies only to this chapter which is devoted to determining the influence of sulfide on the preservation and speciation of inorganic arsenic species.

3. Sodium hypochlorite (NaOCl) added as an oxidant prior to As(Tot) analysis was better than H_2O_2 for the determination of As(Tot) in sulfidic waters preserved with acids or EDTA-HAc.
4. Speciation of As(III) in the EDTA-HAc preserved sample by differential pH (citric acid/citrate buffer, pH 5.0) in the presence of sulfide showed lower values than field speciation anion-exchange resin method when S^{2-} concentration was greater than 0.5 mg/L.
5. For speciation of As(III) and thioarsenite species in the drinking, arsenic containing sulfidic water with no preservative was passed through the Cl-form anion-exchange resin column. The negatively charged thioarsenite species were exchanged onto the resin and uncharged arsenite passed through the column without retention. The concentration of negatively charged thioarsenite species was calculated from the difference between the As(Tot) and As(III) analyses. The adsorbed thioarsenite species could be eluted by NaOCl solution.

Based on the literature and as confirmed in our experiments, the presence of sulfide and arsenic together leads to very complicated thioarsenite chemistry. EDTA-HAc could not effectively preserve the arsenic species when sulfide was present. When sulfide concentration was high (>1.0 mg/L) precipitation of As_2S_3(am) occurred, and only As(Tot) could be measured in preserved samples after the treatment with NaOCl. As(III) and thioarsenite species could be speciated by anion-exchange resin. The best way to measure As(III) and thioarsenite species was to speciate them immediately after collection of sample without any preservative in the field by anion-exchange resin. The difference between As(Tot) and As(III) in the effluent gave the value of thioarsenite species.

Recommendations

Based on the theoretical and experimental results from this project, the following recommendations can be made relative to the analysis and speciation of inorganic arsenic (III)/(V) in ground waters with and without sulfide present

1 Arsenic-containing ground waters that do not contain detectable levels of sulfide may be preserved with EDTA-HAc and immediately speciated in the field with chloride-form resin minicolumns. Samples preserved in this way may be stored at room temperature and analyzed up to 30 days later for As(Tot) and/or speciated again for As(III)/(V) using ion exchange or differential pH hydride generation in the laboratory. For these samples As(V) is retained on the columns and As(III) passes through the column as H_3AsO_3. The relevant equations for non-sulfidic waters are as follows:

$$\text{As(Tot)} = \text{As(III)} + \text{As(V)} \qquad (12)$$
$$\text{As(V)} = \text{As(Tot)} - \text{As(III)} \qquad (13)$$

2 If an arsenic-contaminated ground water to be speciated for As(III)/(V) contains detectable sulfide, it should not be preserved with EDTA-HAc or any mineral acid. Acidifying such a sample will likely precipitate As_2S_3(am) and cause falsely low As(III) and As(Tot) analyses. The unpreserved sample may be analyzed later for As(Tot) by

adding NaOCl (0.02-0.1 M) to oxidize and redissolve arsenic sulfides prior to As(Tot) analysis by hydride generation AAS or alternative method.

3 Sulfide-containing ground waters will contain only As(III) in the form of thioarsenite species and H_3AsO_3, and may be immediately speciated in the field (without preservation) using chloride-form anion resin minicolumns to separate thioarsenite species from H_3AsO_3. The quantification of thioarsenite species may be important because of their lower bioavailability and reduced toxicity compared with arsenite (H_3AsO_3) species.

4 Thioarsenite species concentration may be determined as the difference between As(Tot) and As(III) as measured in the IX eluent of a chloride-form anion column used to speciate an unpreserved sulfidic ground water. The relevant equation are as follows:

$$\text{As(Tot)} = \text{As(Thioarsenite)} + \text{As}(H_3AsO_3) \qquad (14)$$

$$\text{As(Thioarsenite)} = \text{As(Tot)} - \text{As}(H_3AsO_3) \qquad (15)$$

APPENDIX A
ARSENIC PRESERVATION AND FIELD SPECIATION PROCEDURE

Purpose

Speciation of inorganic arsenic is crucial for assessing health implications from arsenic ingestion and for effective removal of arsenic from water. The on-site speciation of inorganic arsenic followed by laboratory quantification will provide the native (point-of-sampling) distribution of arsenic species. This result will be compared with the result of the preserved sample analyzed in the laboratory after collection. This study will evaluate the efficiency of the preservative used to stabilize the arsenic species while they are in transit and in storage before analysis.

Contents of Each Arsenic Speciation Sampling Kit:

1. One Ion-Exchange minicolumn field with 4-mL chloride-form resin.
2. One 250-mL brown bottle marked "C" containing HAc-EDTA solutions for preservation of arsenic.
3. One 30-mL bottle marked "C-1", for collecting the speciation column effluent.
4. One 30-mL syringe for forcing a sample from bottle C through the minicolumn into bottle C-1.
5. Sampling preservation and speciation instruction handout
6. Sample collection data sheet

Procedure for Preservation and Field Speciation Test

Use the following procedure for preservation and field speciation of inorganic arsenic.

(a) Flush all the sample lines thoroughly, and establish a flowing water sample through a flexible tube, typically ¼ inch diameter.
(b) Fill the brown bottle C to a level just below its neck, being careful to avoid turbulence and air entrapment. Don't allow bottle C to overflow because it contains acetic acid and EDTA for pH adjustment and stabilization (Figure A1).
(c) After capping bottle C, mix it thoroughly by inverting it 10 times.
(d) Let bottle C sit for 2-3 minutes and then open the cap to take sample for field speciation study.
(e) Unpack the 30-mL syringe; remove the cap on the syringe tip.
(f) Withdraw 30-mL sample from the bottle C with the syringe. See Photo F1
(g) Grasp the resin minicolumn (The female luer is the top of the column, i.e., the inlet where the syringe is to be connected. See Figure A2) and attached the resin column to the syringe prior to speciation. See Photo F2
(h) Force 30-mL of sample through the resin minicolumn during a period of 15 seconds to 1.5 minutes. See Figure A3 & Photo F3. IMPORTANT: Waste the first 8 mL of effluent from the minicolumn, and collect the last 22 mL effluent in the bottle labeled C-1.

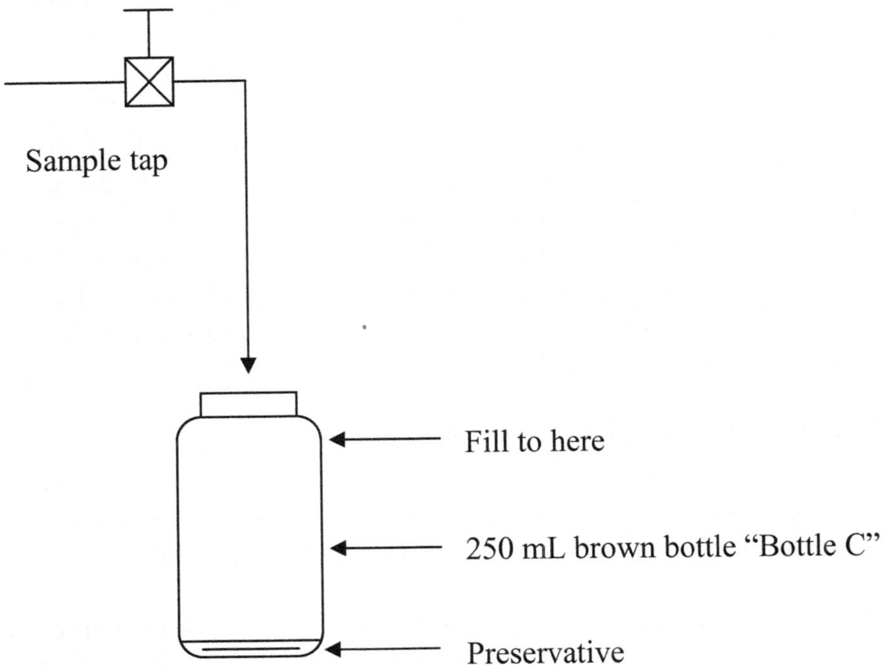

Figure A1. Sample collection into bottle containing preservative

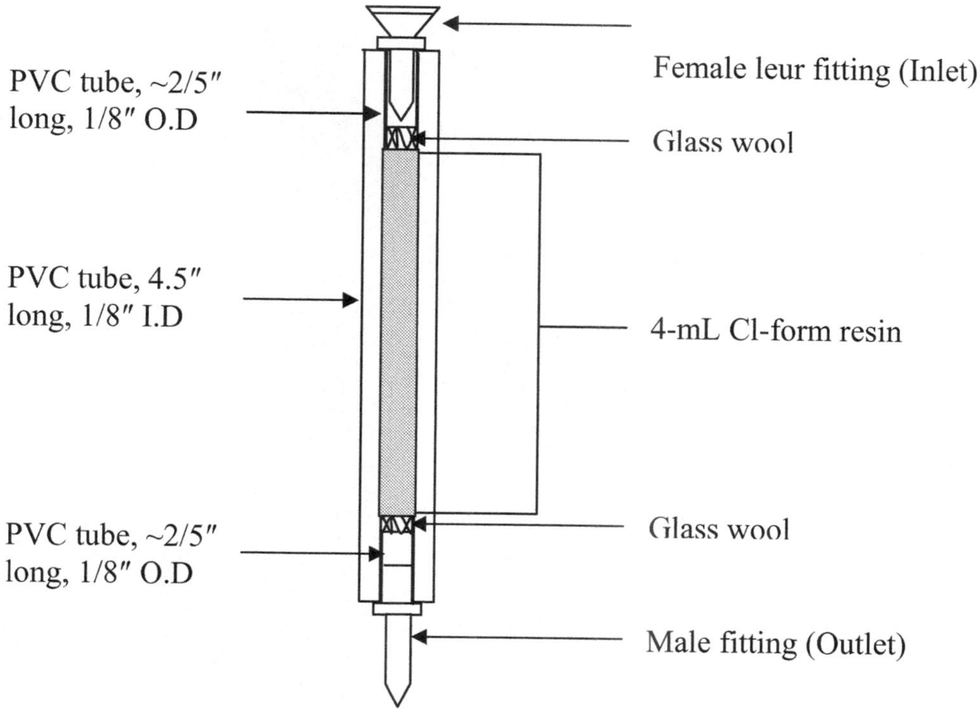

Figure A2. Diagram of resin minicolumn (in the vertical position) showing its components

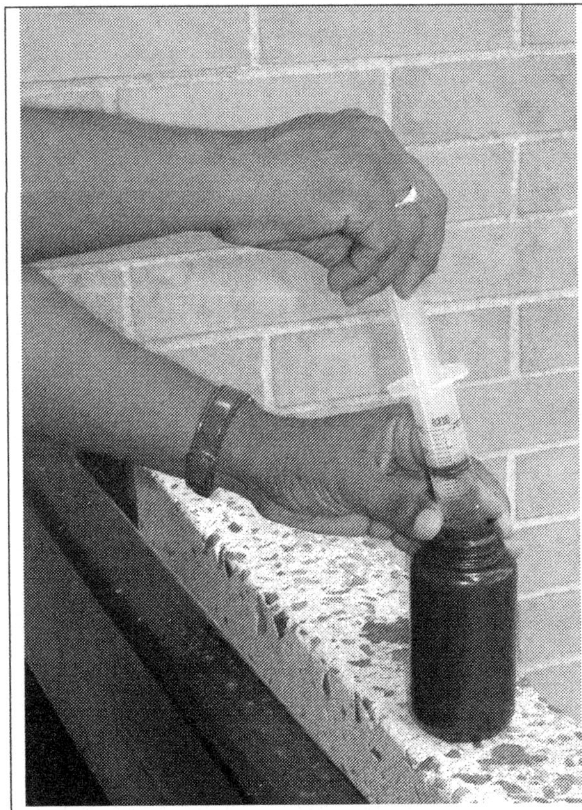

Photo F1. Withdrawing the preserved sample from Bottle C into the 30-mL syringe.

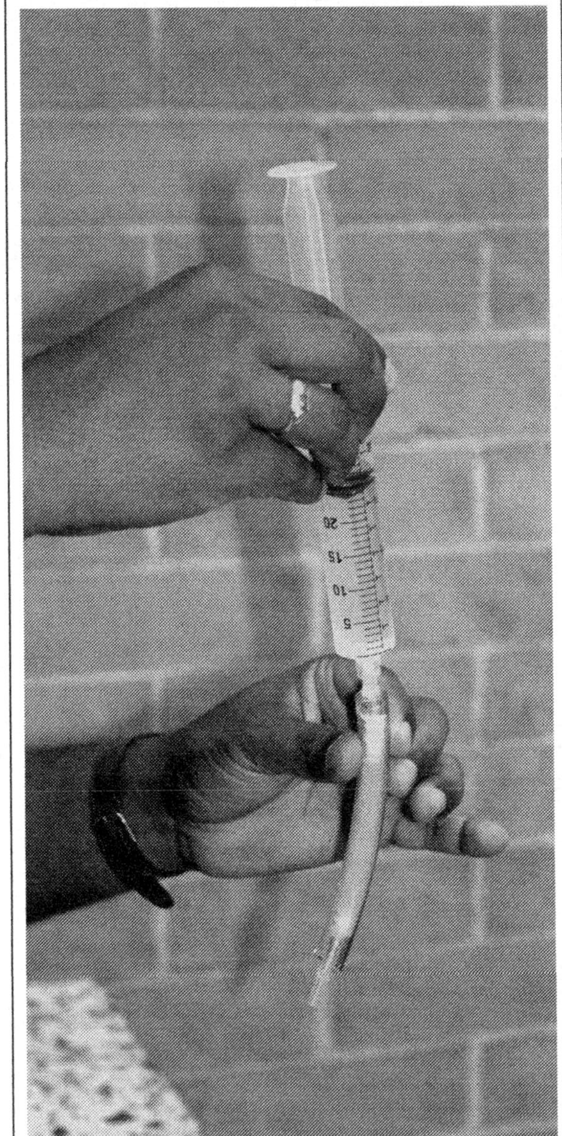

Photo F2. Attaching the resin column to the syringe prior to speciation.

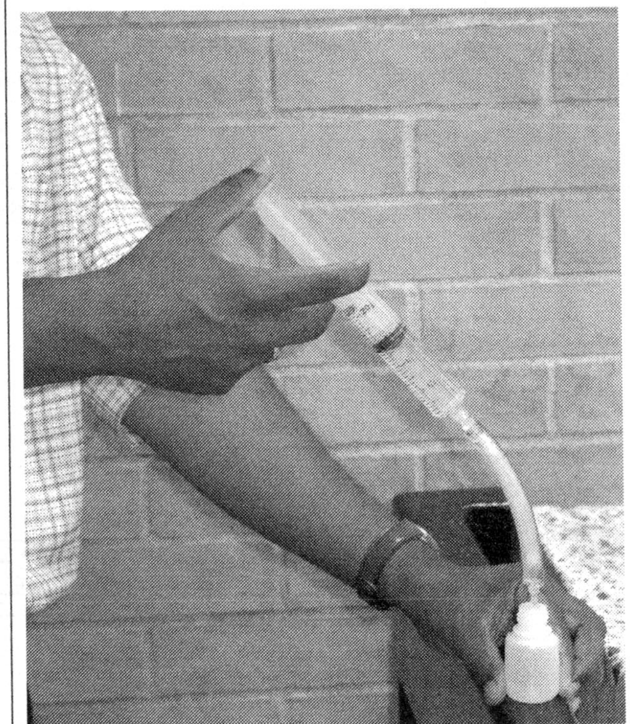

Photo F3. Speciating the preserved sample by forcing it through the resin column.

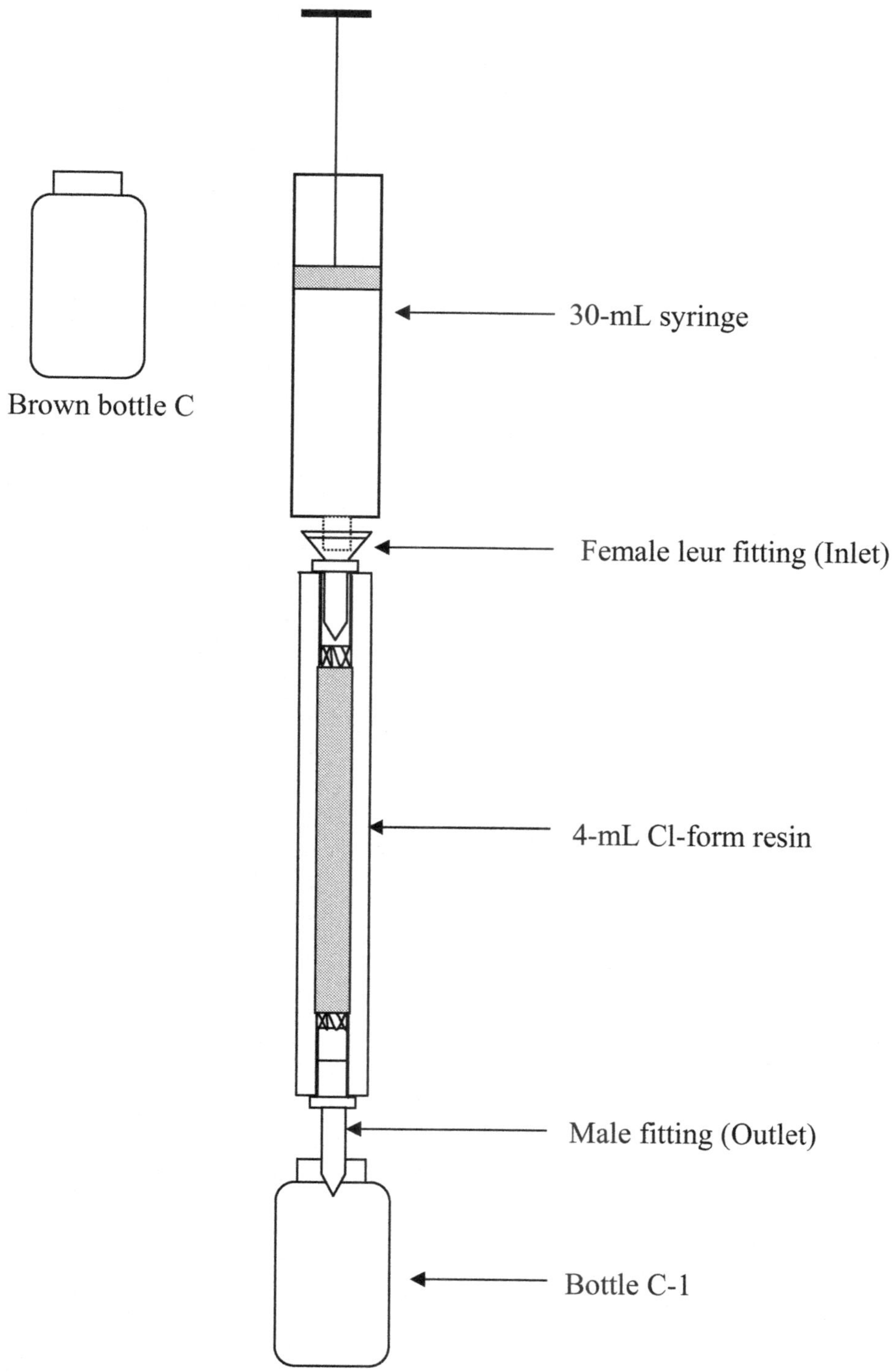

Figure. A3. Field Speciation of the preserved sample.

APPENDIX B
MINNESOTA WELL DATA

Well ID	X	Y	DEPTH
113676	239385	5122947	265
115406	242641	5172494	80
115409	242732	5172341	65
117003	280948	5104272	128
117039	267511	5158557	187
129812	268334	5076613	76
129838	253215	5164848	338
151138	264111	5070492	176
151140	264012	5055766	51
167758	279089	5075062	84
174416	238569	5055980	300
174627	269222	5067617	142
174947	261484	5058883	76
194365	253392	5164485	110
196987	249522	5059755	150
213999	230059	5029059	182
402603	222994	5134599	261
409252	238790	5041991	225
411965	226019	5068004	103
411999	226429	5065165	130
423436	277042	5105437	94
424478	267129	5074111	142
431757	236630	5046916	205
431789	220492	5067932	201
431791	241788	5042825	257
444426	277698	5086434	115
463181	225328	5070179	112
476592	279409	5075971	112
483436	224516	5067293	78
485717	283004	5078036	112
494669	245349	5169579	85
498791	266406	5074196	60
515848	248024	5139126	61
517044	224259	5065132	175
520121	282381	5106238	118
524783	282869	5101550	110
566900	258444	5074279	99
571710	284337	5103131	60
673611	290691	5104738	160

(Continued)

Appendix B (Continued)

Well ID	X	Y	DEPTH
673621	275586	5163074	80
673622	275596	5163083	80
673623	252801	5119055	79
673624	260412	5119215	100
675566	364417	4977679	285

REFERENCES

Aggett, J., and M.R Kriegman. 1987. Preservation of Arsenic(III) and Arsenic(V) in Samples of Sediment Interstitial Water. *Analyst*, 112(2): 153-157.

Al-Sibaai, A. A., and A.G. Fogg.1973. Stability of Dilute Standard Solutions of Antimony, Arsenic, Iron, and Rhenium Used in Colorimetry. *Analyst*. 98(1171): 732-738.

Amarasiriwardena, C.J., Lupoli, N., Potula, V., Korrick, S., and H. Hu. 1998. Determination of the Total Arsenic Concentration in Human Urine by Inductively Coupled Plasma Mass Spectrometry: a Comparison of the Accuracy of Three Analytical Methods. *Analyst*, 123(3): 441-445.

APHA, AWWA, AND WEF (American Public Health Association, American Water Works Association, and Water Environmen Federation). 1992. *Standards Method for the Examination of Water and Wastewater*. 1992. 18th ed. Washington, DC.:APHA.

Balmer, M.; B. Sulzberger.1999. Atrazine Degradation in Irradiated Iron/Oxalate Systems: Effects of pH and Oxalate. *Environ. Sci. Technol.* 33(14): 2418-2424.

Bednar, A. J., J.R. Garbarino, J.F. Ranville, and T.R. Wildeman. 2002. Preserving the Distribution of Inorganic Arsenic Species in Groundwater and Acid Mine Drainage Samples. *Environ. Sci. Technol.*, 36(10): 2213-2218.

Belzile, N. and A. Tessier. 1990. Interactions Between Arsenic and Iron Hydroxides in Lacustrine Sediments. *Geochim. Cosmochim. Acta*, 54(1): 103-109.

Bondietti, G.; J. Sinniger, W. Stumm.1993. The Reactivity of Fe(III)(hydro)oxides: Effects of Ligands in Inhibiting the Dissolution. *Colloid. Surf. A: Physicochem. Eng. Aspects*. 79(2-3): 157-167.

Borho, M., and P. A. Wilderer. 1997. A Reliable Method for Preservation and Determination of Arsenate (III) Concentrations in Groundwater and Water Works. *Jour. water SRT-Aqua*, 46(3) 138-143.

Buxton, G. V., C.L. Greenstock, W.P. Helman, and A.B. Ross. 1988. Critical Review of Rate Constants for Reactions of Hydrated Electrons, Hydrogen Atoms and Hydroxyl Radicals (·OH/·O-) in Aqueous Solution. *Jour. Phys. Chem. Ref. Data*, 17(2): 513-886.

Canfield, D.E. 1989. Reactive iron in marine sediments. *Geochim. Cosmochim. Acta*, 53(5):619-632.

Chakraborti, D., M.M Rahman, K. Paul, U.K. Chowdhury, M.K. Sengupta, D. Lodh, C.R. Chanda, K.C. Saha, and S.C. Mukherjee. 2002. Arsenic Calamity in the Indian Subcontinent What Lessons Have Been Learned? *Talanta*, 58(1):3-22.

Cheam, V., and H. Agemian.1980. Preservation of Inorganic Arsenic Species at Microgram Levels in Water Samples. *Analyst*, 105(1253): 737-743.

Chen, C.J, M.M. Wu, S.S. Lee, J.D. Wang, S.H. Cheng, and H.Y. Wu. 1988. Atherogenicity and Carcinogenicity of High-Arsenic Artesian Well Water: Multiple Risk Factors and Related Malignant Neoplasms of Black Foot Disease. *Arteriosclerosis*, 8(5):452-460.

Cherry, J. A., A.U. Shaikh, D.E. Tallman, and R.V. Nicholson. 1979. Arsenic Species as an Indicator of Redox Conditions in Groundwater. *J. Hydrol.*,43(1-4): 373-392.

Chowdhury U.K, B.K. Biswas, T. Roychowdhury, B.K. Mandal, G. Samanta, G.K. Basu, C.R. Chanda, D. Lodh, K.C Saha, D. Chakraborti, S.C. Mukherjee, S. Roy, S. Kabir, and Q. Quamruzzaman. 2000. Arsenic Groundwater Contamination and Sufferings of People in West Bengal-India and Bangladesh. In *Trace Elements in Man and Animals.* Edited by A.M. Roussel, R.A. Anderson, and A.E. Faviers. New York: Kluwer Academic/Plenum Publishers.

Chowdhury, U.K., B.K. Biswas, T. Roychowdhury, G. Samanta, B.K. Mandal, G.K. Basu, C.R. Chanda, D. Lodh, K.C. Saha, S.K. Mukherjee, S. Roy, S. Kabir, Q.Quamruzzaman, and D. Chakraborti. 2000a. Groundwater Arsenic Contamination in Bangladesh and West Bengal-India. *Environ Health Perspec.*, 108(5):393-397.

Clarke, M.B., and G.R. Helz. 2000. Metal-thiometalate transport of biologically active trace elements in sulfidic environments. 1. Experimental evidence for copper thioarsenite complexing. *Environ. Sci. Technol.*, 34(8):1477-1482.

Cleverley, J.S., L.G. Benning, and B.M. Mountain. 2003. Reaction path modeling in the As-S system: a case study for geothermal As transport. *Appl. Geochem.*, 18(9):1325-1345.

Clifford, D.A, Karori, S., Ghurye, G., and G. Samanta. 2004. Field Speciation Method for Arsenic Inorganic Species. Denver, Colo.: AwwwaRF and Awwa.

Clifford, D.A. 1999. Ion-Exchange and Inorganic Adsorption. In *Water Treatment and Disposal*, 5th Ed., New York: McGraw Hill.

Clifford, D.A., L. Ceber, and S. Chow. 1983. Separation of Arsenic III and Arsenic V by Ion-Exchange. In *Proc. 1983 AWWA Water Quality Technology Conference, Norfolk, VA.* Denver, Colo.: AWWA.

Das D, A. Chatterjee, B.K. Mandal, G. Samanta, D. Chakraborti, and B. Chanda. 1995. Arsenic in Ground Water in six Districts of West Bengal, India: the Biggest Arsenic Calamity in the World, Part 2. Arsenic Concentration in Drinking Water, Hair, Nails, Urine, Skin-Scale and Liver Tissue (Biopsy) of the Affected People. *Analyst*,120(3):917-924.

Daus, B., J. Mattusch, R. Wennrich, and H. Weiss. 2002. Investigation on Stability and Preservation of Arsenic Species in Iron Reach Water Samples. *Talanta,* 58(1): 57-65.

De Vitre, R., N. Belzile, and A. Tessier. 1991. Speciation and Adsorption of Arsenic on Diagenetic Iron Oxyhydroxides. *Limnol. Oceanogr.,* 36(7): 1480-1485.

Dove, P. M., J.D Rimstidt.1985. The Solubility and Stability of Scorodite, FeAsO4.2H2O *Am. Mineral.*, 70(7-8): 838-844.

Eary, L.E. 1992. The solubility of amorphous $As2S_3$ from 25 to 90 ^{0}C. *Geochim. Cosmochim. Acta*, 56(6):2267-2280.

Edwards, M., S. Patel, L. McNeill, H-W. Chen, M. Frey, A.D. Eaton, R.C. Antweiler, and H.E. Taylor. 1998. Considerations in As Analysis and Speciation. *Jour. Am. Wat. Work Ass.*, 90(3): 103-113.

Emett, M.T., and G.H. Khoe. 2001. Photochemical Oxidation of Arsenic by Oxygen and Iron in Acidic Solutions. *Wat. Res.*, 35(3): 649-656.

Farmer, J. G. and M.A. Lovell. 1986. Natural Enrichment of Arsenic in Loch Lomond Sediments. *Geochim. Cosmochim. Acta*, 50(9): 2059-2067.

Feldman, C. 1979. Improvements in the Arsine Accumulation-helium Glow Detector Procedure for Determining Traces of Arsenic. *Anal. Chem.*,51(6): 664-669.

Ferguson, J. F. and J. Gavis. 1972. A Review of the Arsenic Cycle in Natural Water. *Wat. Res.*, 6(11): 1259:1274.

Ferreccio, C., C. González, V. Milosavjlevic, G. Marshall, A.M. Sancha, and A.H. Smith. 2000. Lung Cancer and Arsenic Concentrations in Drinking Water in Chile. *Epidemiol,* 11(6): 673-679.

Ficklin, W.H. 1983. Separation of arsenic (III) and arsenic(V) in ground waters by ion exchange, *Talanta*, 30(5): 371-373.

Floroiu, R. M.; A.P. Davis, A.Torrents. 2004. Kinetics and Mechanism of As_2S_3(am) Dissolution under N_2. *Environ. Sci. Technol.*, 38(4): 1031-1037.

Frank, R., and H. Rau. 1990. Photochemical Transformation in Aqueous Solution and Possible Environmental Fate of Ethylenediaminetetraacetic acid (EDTA). *Ecotoxicol. Environ. Saf.*, 19(1): 55-63.

Gallagher, P. A., C.A. Schwegel, X. Wei, and J.T. Creed. 2001. Speciation and Preservation of Inorganic Arsenic in Drinking Water Sources Using EDTA With IC Separation and ICP-MS Detection. *Jour. Environ. Monit.*, 3(4): 371-376.

Guha Mazumder, D.N, R. Haque, N. Ghosh, B.K. De, A. Santra, D. Chakraborty, and A.H. Smith. 1998. Arsenic Levels in Drinking Water and the Prevalence of Skin Lesions in West Bengal. *India. Int J Epidemiol.*, 27(5):871-877.

Hall, G.E.M., J.C. Pelchat, and G. Gauthier. 1999. Stability of Inorganic Arsenic (III) and Arsenic (V) in Water Samples. *Jour. Anal. At. Spectrom.*, 14(2): 205-213.

Helz, G.R., J.A. Tossell, J.M. Charnock, R.A.D. Pattrick, D. J. Vaughan, and C.D. Garner. 1995. Oligomerixation in As(III) Sulfide Solutions: Theoretical Constraints and Spectroscopic evidence. *Geochim. Cosmochim. Acta*, 59(22): 4591-4604.

Hug, S.J., L. Canonica, M. Wegelin, D. Gechter, and U.V. Gunten. 2001. Solar Oxidation and Removal of Arsenic at Circumneutral pH in Iron Containing Waters. *Environ. Sci. Technol.*, 35(10): 2114-2121.

Huie, R. E., L.C.T. Shoute, and P. Neta.1991. Temperature Dependence of the Rate Constants for Reactions of the Carbonate Radical With Organic and Inorganic Reductants. *Int. J. Chem. Kinet.* 23(6): 541-552.

Irgolic, K.J. 1994. Determination of Total Arsenic and Arsenic Compounds in Drinking Water. In *Arsenic Exposure and Health.* Edited by W.R. Chappell, C.O. Abernathy, C.R. Cothern. Northwood: Science and Technology Letters.

Jay J.A., N.K. Blute, H.F. Hemond, and J.L. Durant. 2004. Arsenic-sulfides confound anion-exchange resin speciation of aqueous arsenic. *Wat. Res.*, 38(5): 1155-1158.

Karametaxas, G., S.J. Hug, and B. Sulzberger. 1995. Photodegradation of EDTA in the Presence of Lepidocrocite. *Environ. Sci. Technol.* 29(12): 2992-3000.

Kim, M.J. 2001. Separation of Inorganic Arsenic Species in Groundwater Using ion Exchange Method. *Bull. Environ. Contam. Toxicol.*, 67(1): 46-51.

Kim, M.-J., and J. Nriagu. 2000. Oxidation of Arsenite in Groundwater Using Ozone and Oxygen. *Science of the Total Environ.*, 247(1): 71-79.

Kinniburgh, D. G. and W. Kosmus. 2002. Arsenic Contamination in Groundwater: Some Analytical Considerations. *Talanta*, 58(1):165-180.

Klaning, U. K., B.H.J. Bielski. K. Sehested.1989. Arsenic (IV). A Pulse-Radiolysis Study. *Inorganic Chemistry*, 28(14): 2717-2724.

Krupp, R.E. 1990. Comments on "As(III) and Sb(III) sulfide complexes: an evaluation of stoichiometry and stability from existing experimental data" by N. F. Spycher and M. H. Reed. *Geochim. Cosmochim. Acta*, 54(11):3239-3240.

Le, X.C., S. Yalcin, and M, Ma. 2000. Speciation of submicrogram per liter levels of arsenic in water: On-site species separation integrated with sample collection. *Environ. Sci. Technol.*, 34(11): 2342-2347.

Lee, H., and W. Choi. 2002. Photocatalytic Oxidation of Arsenite in TiO_2 Suspension: Kinetics and Mechanism. *Environ. Sci. Technol.*, 36(17): 3872-3878.

Lindsay, W. L. 1979. *Chemical Equilibria in soil*, 1st ed., New York: Wiley.

Litter, M. I., E.C. Baumgartner, G.A. Urrutia, and M. Blesa.1991. Photodissolution of Iron oxides 3. Interplay of Photochemical and Thermal Processes in Maghemite/Carboxylic Acid Systems. *Environ. Sci. Technol.* 25(11): 1907-1913.

Lytle, D. and V. Snoeyink. 2002. Effect of Ortho- and Polyphosphates on the Properties of Iron Particles and Suspensions. *Jour. AWWA*. 94(10): 87-99.

Meng, X., C. Jing, and G.P. Korfiatis. 2003. A Review of Redox TraNSF 2ormation of arsenic in Aquatic Environments. In *Biogeochemistry of Environmentally Important Trace Elements*. Edited by Y. Cai and O.C. Braids. Washington D.C.: ACS Symposium Series 835, American Chemical Society.

Migdisov, A.A., A.E. Williams-Jones, L.Z. Lakshtanov, and Y.V. Alekhin. 2002. Estimates of the second dissociation constant of H2S from the surface sulfidation of crystalline sulfur. *Geochim. Cosmochim. Acta*, 66(8): 1713-1725.

Millero, F. J., and S. Sotolongo.1989. The Oxidation of Iron(II) With Hydrogen Peroxide in Seawater. *Geochim. Cosmochim. Acta*. 53(8): 1867-1873.

Millero, F. J., S. Sotolongo, and M. Izaguirre. 1987. The Oxidation Kinetics of Iron(II) in Seawater. *Geochim. Cosmochim. Acta* 51(4): 793-801.

Mironova, G.D., A.V. Zotov, and N.I. Gulko. 1984. The solubility of orpiment in sulfidic solution at 25-150 °C and the stability of arsenic sulfide complexes. *Geochem. Int.*, 21(1):53-59.

Mironova, G.D., A.V. Zotov. 1980. Solubility studies of the stability of As(III) sulfide complexes. *Geochem. Int.*, 17(1):46-54.

Moore, J. N., W.H. Ficklin, and C. Johns. 1988. Partitioning of Arsenic and Metals in Reducing Sulfidic Sediments. *Environ. Sci. Technol.*, 22(4): 432-437.

Nickson, R, J. McArthur, W. Burgess, K.M. Ahmed, P. Ravenscroft, and M. Rahman. 1998. Arsenic Poisoning of Bangladesh Groundwater. *Nature*,395(6700):338.

NRC (National Research Council, Subcommittee on Arsenic in Drinking Water). 1999. Arsenic in drinking water. National Academy Press, Washington, DC.

NRDC (National Resources Defense Council). 2000. Arsenic and old Laws. A Scientific and Public Health Analysis of Arsenic Occurrence in Drinking Water, its Health Effects, and EPA's Outdated Arsenic tap Water Standard. (URL: www.nrdc.org/water/drinking/arsenic/chap1.asp).

O'Day, P.A., D. Vlassopoulos, R. Root, and N. Rivera. 2004. The influence of sulfur and iron on dissolved arsenic concentrations in the shallow subsurface under changing redox conditions. *PNAS*, 101(38):13703-13708.

Oscarson, D. W., P.M. Huang, C. Defosse, and A. Herbillon. 1981. Oxidative Power of Manganese(IV) and Iron(III) Oxides With Respect to Arsenic(III) in Terrestrial and Aquatic Environments. *Nature*, 291(5810): 50-51.

Palacios, M. A., M. Gomez, C. Camara, and M.A. Lopez. 1997. Stability Studies of arsenate, Monomethylarsonate, Dimethylarsinate, Arsenobetaine and arsenochlorine in Deionised Water, Urine and Clean-up dry Residue From Urine Samples and Determination by Liquid Chromatography With Microwave-Assisted Oxidation –Hydride Generation Atomic Absorption Spectrometric Detection. *Anal. Chim. Acta*, 340(1-3): 209-220.

Pettine, M., L. Campanella, and F.J. Millero.1999. Arsenite Oxidation by H_2 O_2 in Aqueous Solutions *Geochim. Cosmochim. Acta*, 63(18): 2727-2735.

Pierce, M.L., and C.B. Moore. 1982. Adsorption of Arsenite and Arsenate on Amorphous Iron Hydroxide. *Wat. Res.*, 16(7):1247-1253.

Pokrovski, G., R. Gout, J. Schott, A.V. Zotov, and J. Harrichoury. 1996. Thermodynamic properties and stochiometry of As(III) hydroxide complexes at hydrothermal conditions. *Geochim. Cosmochim. Acta*, 60(5): 737-749.

Rader K.J, P.M. Dombrowski, K.J. Farley, J.D. Mahony, and D.M. Di Toro. 2004. Effect of thioarsenite formation on arsenic(III) toxicity. *Environ. Toxicol. Chem.*, 23(7):1649-1654.

Rochette, E.A., B.C. Bostick, G. Li, and S. Fendorf. 2000. Kinetics of Arsenate Reduction by Dissolve Sulfide. Environ. Sci. Technol., 34(22): 4714-4720.

Roy Chowdhury, T., G.K. Basu, B.K. Mandal., B.K. Biswas, G. Samanta, U.K. Chowdhury, C.R. Chanda., D. Lodh, S.L. Roy, K.C. Saha, S. Roy, S. Kabir, Q. Quamruzzaman and D. Chakraborti. 1999. Arsenic Poisoning in the Ganges Delta. *Nature*, 401(6753): 545-546.

Schecher, W.D., and D.C. McAvoy. 2001. MINEQL+: A Chemical Equilibrium Modeling System, V. 4.5 for Windows. Environmental Research Software, Hallowell, ME.

Shock, E.L., D.C. Sassani, M. Willis, and D.A. Sverjensky, 1997. Inorganic species in geological fluids: Correlations among standard molal thermodynamic properties of aqueous ions and hydroxide complexes. *Geochim.Cosmochim. Acta*, 61(5):907-950.

Smedley, P.L. and D.G. Kinniburgh. 2002. A Review of Source, Behavior and Distribution of Arsenic in Natural Waters. *Appl. Geochem.*, 17(5): 517-568.

Smieja, J.A. and R.T. Wilkin. 2003. Preservation of Sulfidic Waters Containing Dissolved As(III). *Jour. Environ. Monit.*, 5(6): 913-916.

Smith, R. M., and A.E. Martell.1979. *Critical Stability Constants*. New York: Plenum.

Spycher, N.F., and M.H. Reed. 1989. As(III) and Sb(III) sulfide complexes: an evaluation of stoichiometry and stability from existing experimental data, Geochim. Cosmochim. Acta, 53(9):2185-2194.

Stuglik, Z.and Z.P. Zagorski. 1975. Radiolytic Oxidation of Iron(II) in a Neutral Aqueous Solution. *Radiat. Radioanal. Lett.* 22(4): 257-262.

Stumm, W., and J.J. Morgan. 1996. *Aquatic Chemistry, Chemical Equilibria and Rates in Natural Waters, 3rd ed.*; New York: John Wiley & Son.

Sulzberger, B.. *In Chemistry of the Solid-Water Interface*; Stumm, W., Ed.; John Wiley and Sons: New York, 1992; p 428.

Takamatsu, T., M. Kawashima, and M. Koyama. 1985. The Role of Mn^{2+}-Rich Hydrous Manganese Oxides in the Accumulation of Arsenic in Lake Sediments. *Wat. Res.*, 19(8): 1029-1032.

Tallman, D. E., and A.U. Shaikh.1980. Redox Stability of Inorganic Arsenic(III) and Arsenic(V) in Aquous Solution. *Anal. Chem.*, 52(1): 196-199.

Tossel, J.A. 1996. Theoretical studies on As and Sb sulfide molecules, in Dyar, M. D., WcCammon, C., Schaefer, M. W. (Eds), Mineral Spectroscopy: A tribute to Roger G. Burns., *Geochem. Soc. Spec. Publ.*, 5:9-21.

Tossel, J.A. 1997. Theoretical studies on arsenic oxide and hydroxide species inminerals and in aqueous solution. *Geochim.Cosmochim. Acta*, 61(8):1613-1623.

Tossell, J.A. 2000. Metal-Thiometalate Transport of Biologically Active Trace Elements in Sulfidic Environments. 2. Theoretical Evidence for Copper Thioarsenite Complexing. *Environ. Sci. Technol.*, 34(8): 1483-1488.

Turner, R. 1981. Oxidation State of Arsenic in Coal ash Leachate. *Environ. Sci. Technol.*, 15(9): 1062-1066.

U.S. EPA (United States Environmental Protection Agency). 2001. EPA announces arsenic standard for drinking water of 19 parts per billion, EPA headquarters press release, Washington, DC, 10/31/2001.

Van Elteren, J. T., J. Hoegee, E.E. Van der Hoek, H.A. Das, C.L. de Ligny, and J. Agterdenbos. 1991. Preservation of Arsenic(III) and Arsenic(V) in Some Water Samples. *J. Radioanal. Nucl. Chem. Lett.*, 154(5): 343-355.

Wagman, D.D., W.H. Evans, V.B. Parker, R.H. Schumm, I. Halow, S.M. Bailey, K.L Churney and R.L. Nuttall. 1982. The NBS tables of chemical thermodynamic properties. *Jour. Phys. Chem. Ref. Data*. 11:1-392.

Walling, C. 1975. Fenton's Reagent Revisited. *Acc. Chem. Res.*, 8(4): 125-131.

Watt, C and X.C. Le. 2003. Arsenic Speciation in Natural Waters. In *Biogeochemistry of Environmentally Important Trace Elements*. Edited by Y. Cai and O.C. Braids. Washington D.C.: ACS Symposium Series 835, American Chemical Society.

Webster, J. G. 1990. The Solubility of AS_2S_3 and Speciation of As in Dilute and Sulphide-Bearing Fluids at 25 and 90 °C. *Geochim. Cosmochim. Acta*, 54(5): 1009-1017.

Wilkie, J. A., and J.G. Hering. 1998. Rapid Oxidation of Geothermal Arsenic (III) in Streamwaters of the Eastern Sierra Nevada. *Environ. Sci. Technol.*, 32(5): 657-662.

Wilkin, R.T., D. Wallschläger, and R.G. Ford. 2003. Speciation of arsenic in sulfidic waters. *Geochem. Trans.*, 4(1):1-7.

Wilmot, P.D., K. Cadee, J.J. Katinic, and B.V. Kavanagh. 1988. Kinetics of sulfide oxidation by dissolved oxygen. *Jour., WPCF.*, 60(7):1264-1270.

Wu, M. 2001. Adsorption of Arsenic onto Activated alumina: the Effects of pH, Silica and hardness. Master's thesis. University of Houston, Houston.

ABBREVIATIONS

µg/L	microgram/liter
d	day
DO	dissolve oxygen
EDTA	ethylenediaminetetraacetic acid
FI-HG-AAS	flow injection hydride generation atomic absorption spectrometer
h	hour
HAc	acetic acid
ICP-MS	inductively coupled plasma-mass spectrometer
M	molar
MCL	maximum contaminant level
mg/L	milligram/liter
mM	millimolar
mV	millvolt
nM	nano meter
NRC	national Research Council
NRDC	natural Resources Defense Council
PET	polyethyleneterephthalate
PP	polypropylene
ppm	parts per million
RSD	relative standard deviation
TDS	total dissolve solid
USEPA	U.S. Environmental Protection Agency
USGS	U.S. Geological Survey
UV	ultraviolet
w/m^2	watt/square meter

IWA Publishing's authorised EU representative for General Product Safety Regulations is Diane D'Arras, 15 rue Duret, 75116 Paris, France, e-mail: safety@iwap.co.uk.

Printed and bound by CPI Group (UK) Ltd, Croydon, CR0 4YY
17/09/2026
02218799-0018